Solve It! 3rd:

Problem-Solving Strategies

Authors
David Mitchell
Myrna Mitchell
Michelle Pauls
Dave Youngs

Editors
Michelle Pauls
Betty Cordel

Illustrators
Dawn DonDiego
Broc Heasley
Reneé Mason
Margo Pocock
Dave Scholotterback
Brenda Wood

Desktop Publisher
Roxanne Williams

Education Foundation

This book contains materials developed by the AIMS Education Foundation. **AIMS** (**A**ctivities **I**ntegrating **M**athematics and **S**cience) began in 1981 with a grant from the National Science Foundation. The non-profit AIMS Education Foundation publishes hands-on instructional materials (books and the quarterly magazine) that integrate curricular disciplines such as mathematics, science, language arts, and social studies. The Foundation sponsors a national program of professional development through which educators may gain both an understanding of the AIMS philosophy and expertise in teaching by integrated, hands-on methods.

ISBN: **978-1-932093-18-6**

Printed in the United States of America

I Hear and

I Forget,

I See and

I Remember,

I Do and

I Understand.

-Chinese Proverb

Solve It! 3rd:

Problem-Solving Strategies

Introduction

Solve It! 3rd: Problem-Solving Strategies is a collection of activities designed to introduce young children to nine problem-solving strategies. The tasks included will engage students in active, hands-on investigations that allow them to apply their number, computation, geometry, data organization, and algebra skills in problem-solving settings.

It can be difficult for teachers to shift from teaching math facts and procedures to teaching with an emphasis on mathematical processes and thinking skills. One might ask why problem solving should be taught at all. The most obvious reason is that it is part of most mathematics curricula. However, it is also an interesting and enjoyable way to learn mathematics; it encourages collaborative learning, and it is a great way for students to practice mathematical skills. This in turn leads to better conceptual understanding—an understanding that allows students to remember skills and be able to apply them in different contexts.

Introducing students to the nine strategies included in this book gives them a toolbox of problem-solving methods that they can draw from when approaching problems. Different students might approach the same problem in a variety of ways, some more sophisticated than others. Hopefully, every child can find one approach that he or she can use to solve the problems that you present. Over time, and from discussing what other children have done, students will develop and extend the range of strategies at their disposal.

It is our hope that you will use the problems in this book to enrich your classroom environment by allowing your students to truly experience problem solving. This means resisting the urge to give answers; allowing your students to struggle, and even be frustrated; focusing on the process rather than the product; and providing multiple, repeated opportunities to practice different strategies. Doing this can develop a classroom full of confident problem solvers well equipped to solve problems, both in and out of mathematics, for years to come.

Problem-Solving Strategies

 **Use Manipulatives**

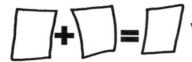 **Write a Number Sentence**

 **Draw out the Problem**

 **Guess and Check**

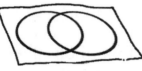 **Organize the Information**

 **Look for Patterns**

 **Use Logical Thinking**

 **Work Backwards**

 **Wish for an Easier Problem**

Activities / Strategies	Use Manipulatives	Write a Number Sentence	Draw out the Problem	Guess and Check	Organize the Information	Look for Patterns	Use Logical Thinking	Work Backwards	Wish for an Easier Problem
Celebrating Combinations	X								
In So Many Words	X								
Flipping Over Symmetry	X								
It All Adds Up		X			X				
That's Sum Name!		X							
What is the One?	X	X							
Schmoos 'n' Goos	X		X						
Sawing Logs			X						X
Picturing Clues			X						
The Lily Pad Hop			X						
Pumpkin Patches	X			X					
Balance Bazaar	X			X					
Score Keepers	X			X					
Probably Bears	X				X				
Bear Hunt					X		X		
Cash Combos					X				
Pattern Play	X					X	X		
Snappy Patterns	X					X			
What's My Rule?	X				X	X			
Fabulous Four-sum	X				X	X			
Apple Arrays	X			X			X		
Historical Logic	X						X		
Clue Me In							X		
What's the Question?	X	X						X	
What's the Scoop?	X							X	
Time Tellers								X	
Blocking Out Fractions	X							X	
Tallying Toothpick Triangles	X				X	X			X
One Step at a Time					X				X
Cube Face Estimation									X

Problem-Solving Strategies
Use Manipulatives

Sometimes it is helpful to use objects when solving a problem. These objects can represent the parts of the problem. Seeing the parts can help you understand how to find the answer. Anything can be a manipulative. You can use paper clips, pattern blocks, Unifix cubes, or even pieces of paper.

Celebrating Combinations

Topic
Combinations

Key Question
How many ways can you dress the snowman, dress the elf and decorate the gingerbread man?

Learning Goals
Student will:
1. use manipulatives to determine all possible combinations of snowmen, elves, gingerbread men, and
2. record their solutions.

Guiding Document
NCTM Standards 2000*
- *Build new mathematical knowledge through problem solving*
- *Solve problems that arise in mathematics and in other contexts*

Math
Combinations
Problem solving

Integrated Processes
Observing
Comparing and contrasting
Recording

Problem-Solving Strategy
Use manipulatives

Materials
Student pages
Card stock
Scissors
Colored pencils or crayons

Background Information
The concept of combinations is one we deal with on a daily basis, likely without giving it a second thought. Which pair of shoes to wear with which pair of pants? Which flavors of ice cream to pick for a two-scoop sundae? Which side dish to order with our main course? The familiarity of these decisions can often cause us to miss the underlying mathematical principles that are embedded there. How many outfits *could* you create with your shirts, shoes, and pants? What are all of the *possible* two-scoop sundaes that can be made with the local ice cream shop's flavors? How many *different* dinners can you order simply by changing the side dish or main course? It is these questions, at a much more basic level, that students will be exploring in this activity.

Management
1. There are three separate challenges presented in this activity, each with a differing level of difficulty. Begin with the snowman, and work up to the elf.
2. Copy the manipulatives for each problem onto cardstock for durability and ease of handling.

Procedure
1. Distribute the first student page, scissors, and crayons or colored pencils to each student.
2. Have them color and cut out the options for dressing the snowman.
3. Instruct the students to use the manipulatives to help them discover all of the possible ways to dress the snowman. (In each case, the snowman must have two accessories, i.e., he cannot wear only a scarf or only a hat.)
4. Have them record each solution they discover by drawing it on one of the small snowmen at the bottom of the page. (There are more spaces to record than there are possible solutions.)
5. Repeat this process with the gingerbread men and the elves.
6. Once the time of discovery is over, go over the solutions and discuss combinations as a class. Have students think of real-life examples where they use combinations. (You may want to bring in the menu from a restaurant or a couple of different outfits to make the discussion more concrete.)

Connecting Learning
1. How many ways are there to dress the snowman? [4]
2. How does this number compare to the number of ways to dress the elf or decorate the gingerbread man? [There are 8 ways to dress the elf and 9 ways to dress the gingerbread man.]
3. Does it surprise you that there are more combinations for the gingerbread man? Why or why not?
4. Which of the three challenges was the most difficult for you? Why?
5. How did you know when you had found all of the solutions?
6. When do you use combinations in your everyday life?

Extension
Have students create one additional accessory for each of the three problems and determine the number of combinations with this addition.

*   Reprinted with permission from *Principles and Standards for School Mathematics*, 2000 by the National Council of Teachers of Mathematics. All rights reserved.

Celebrating Combinations

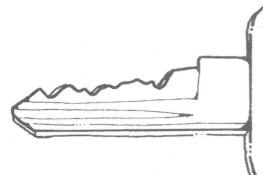

Key Question
How many ways can you dress the snowman, dress the elf decorate the gingerbread man?

Learning Goals

Students will:

1. use manipulatives to determine all possible combinations of snowmen, elves, gingerbread men and
2. record their solutions.

Celebrating Combinations

How many different ways can you dress the snowman when you have
- a top hat or a baseball cap and
- a scarf or buttons?

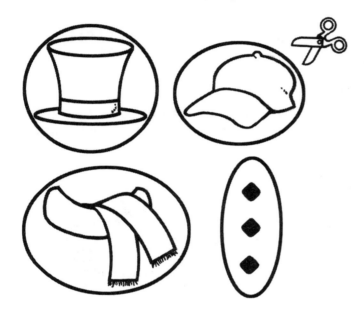

Draw each solution you can find.

Celebrating Combinations

How many different ways can you dress the elf when you have two shirts, two hats, and two pairs of shoes?

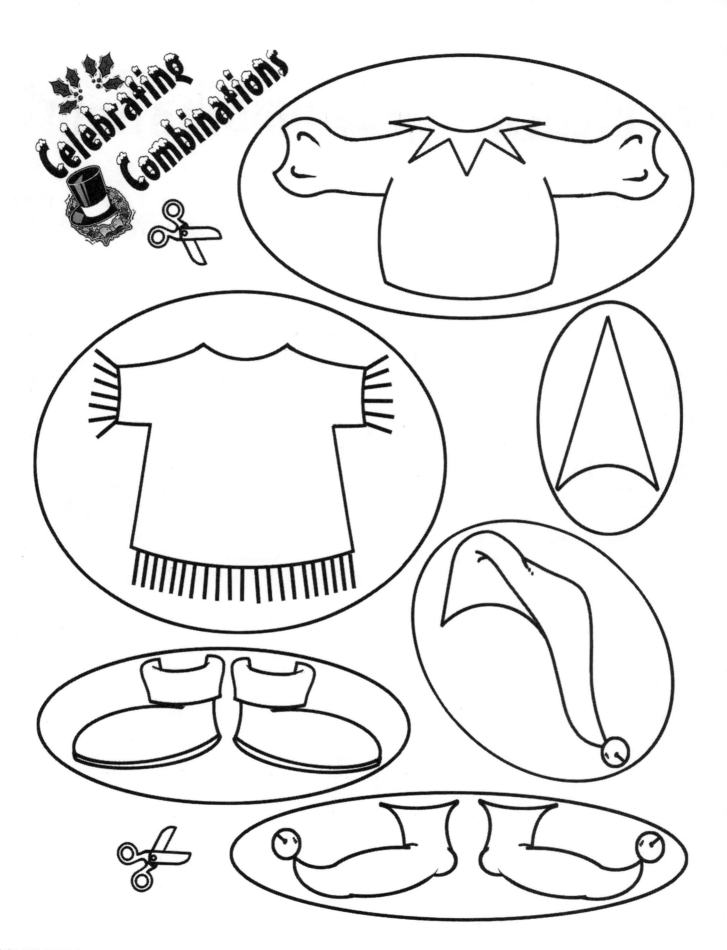

Celebrating Combinations

Draw each solution you find.

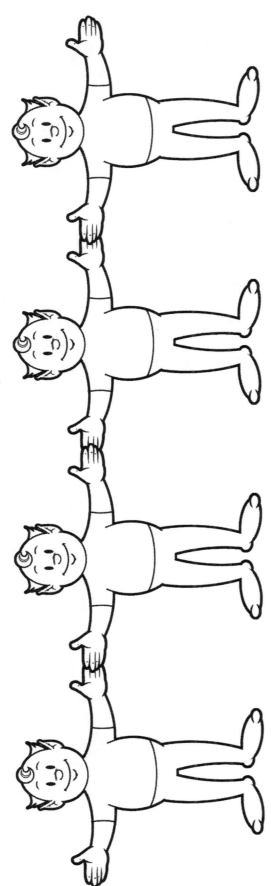

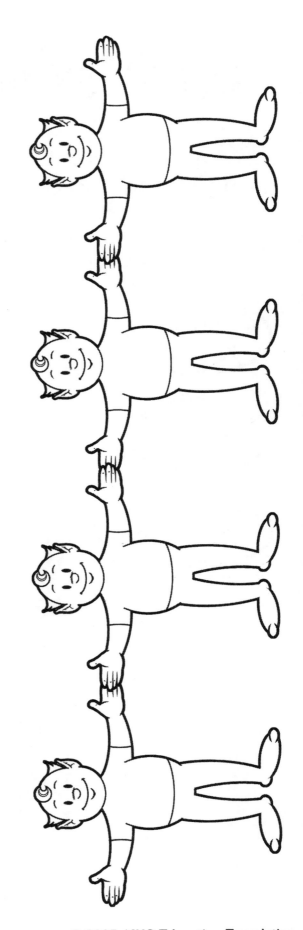

Celebrating Combinations

How many different ways can you make the gingerbread man's mouth and buttons when you have chocolate candies, gumdrops, and raisins?

Choose a mouth

Choose a set of buttons

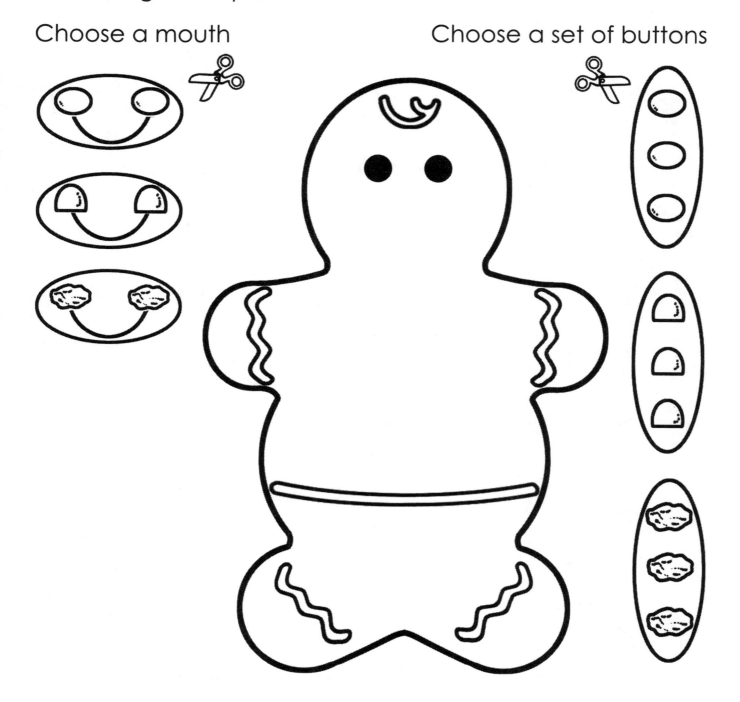

Celebrating Combinations

Draw each solution you find.

Celebrating Combinations

Connecting Learning

1. How many ways are there to dress the snowman?

2. How does this number compare to the number of ways to dress the elf or decorate the gingerbread man?

3. Does it surprise you that there are more combinations for the gingerbread man? Why or why not?

4. Which of the three challenges was the most difficult for you? Why?

5. How did you know when you had found all of the solutions?

6. When do you use combinations in your everyday life?

In So Many WORDS

Topic
Fractions

Key Questions
1. How many fractional words can you find within a large word?
2. How many different words can you make using the same set of letters, and what fraction of the letters do they use?

Learning Goals
Students will:
1. find all of the words hidden within a larger word that use only a fraction of the original letters—without rearranging the order, and
2. use the letters from the original words to create as many words as possible by arranging the letters in any order.

Guiding Documents
Project 2061 Benchmark
- *Use whole numbers and simple, everyday fractions in ordering, counting, identifying, measuring, and describing things and experiences.*

NCTM Standards 2000*
- *Understand and represent commonly used fractions such as ¼, ⅓, and ½*
- *Build new mathematical knowledge through problem solving*
- *Solve problems that arise in mathematics and in other contexts*

Math
Fractions
Problem solving

Integrated Processes
Observing
Comparing and contrasting
Recording

Problem-Solving Strategy
Use manipulatives

Materials
Card stock
Scissors
Crayons or colored pencils
Student pages
Chart paper, optional

Background Information
In this activity, students will have the opportunity to explore fractions in an open-ended format through the use of words. This allows them to experience fractions in a way that they will not typically encounter and is a good way to assess their understanding of the concept. They will first use their problem-solving skills to find small words within larger words and identify the fractional part of the larger word that is made up by the small word. Next, they will be allowed to create words from a specific set of letters and identify what fraction of the letters they used for each word. This activity has the opportunity for many extensions and the ability to integrate nicely with language arts or other subject areas.

Management
1. This activity is divided into two parts, and each part has varying levels of difficulty. Depending on the needs and abilities of your students, you may wish to do only one of these parts.
2. Copy the page of letter cards onto card stock and cut them out for the students. One page contains two sets of letter cards, and each student will need his or her own set. Depending on the words you use, you may or may not need all of the letters provided. (One blank card is included with each set for any additional letters desired.)
3. Words used in this activity were chosen because they have many possibilities for fractional words. Some of them may not be in your students' vocabularies, providing an excellent opportunity for students to learn new words.

Procedure
Part One
1. Tell students that they are going to be using letter cards to make words and look for hidden words within those words. Distribute one set of letter cards to each student.
2. Have the students use the letters M, A, T and E to spell *mate*. All other letter cards should be set aside. Go over some meanings of of *mate* if students are unfamiliar with the word. [a Navy officer, the match to something, an animal or person's partner, etc.]
3. Ask the class if they can find a word hidden inside *mate* that only uses half of the letters. Inform students that they may not rearrange the order of the letters. [ma or at] Discuss why *at*

and *ma* use half of the letters in *mate*. [They use two out of four letters, which is half.]

4. Challenge students to find words hidden inside *mate* that use more than half of the letters. [ate, mat] Have students verbalize the fact that these words use three out of four letters.

5. Ask if there are any words that use less than half of the letters. [a] Have students verbalize the fact that this word uses one out of four letters.

6. Using an overhead transparency of the recording page, record each of the fractional words within *mate* as illustrated. (You will need to write the word MATE in each of the spaces.) Color the box(es) containing each letter used for a given fractional word, and write the number of letters used on the top half of the fraction to the right of the boxes.

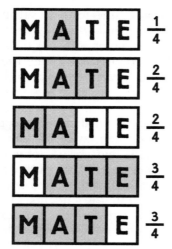

7. Explain that the bottom number tells you the total number of letters (4), and the top number tells you how many letters were used for each word (1, 2, or 3). Have students use the words *one-fourth*, *two-fourths*, and *three-fourths* to describe the different fractional words.

8. Have students get into small groups of two to four and use the letters S, E, A and T to spell *seat*. Challenge students to find all of the hidden words in *seat* without rearranging the letters.

9. Distribute the first student recording page and crayons or colored pencils, and assist groups in recording the solutions they discovered. Have groups share their solutions and discuss the process.

10. Record the words discovered by the groups and the fractional value of each word on the chalkboard or a piece of chart paper.

11. Repeat this process as desired with additional words. Other words with multiple fractional word possibilities that can be made using the letter cards provided are TONE, PANT, and SPAT. Alternately, develop your own words based on vocabulary lists or current topics of study and have students make any additional letter cards necessary.

12. If appropriate, move on to six-, or even eight-letter words. The suggested words using the letter cards are HEARTS, STARED, TEASED, and SHOWERED. Only one recording page is provided for both six- and eight-letter words, so you may wish to modify it to suit your purposes or create your own. Always discuss the fractional words discovered and record them for the whole class to see.

Part Two

1. Choose one of the words that students have been working with and explain that the new challenge is to find fractional words, but now there is no restriction on the order of the letters. If desired, each group can be given a different word. Students should be informed that they may use some or all of the letters in the original word to form new words.

2. Give groups additional recording pages and have them write the word they are using at the top of the page. To record the new words discovered, students should write the letters in the spaces and record the number of letters used. They do not need to color any boxes or record any unused letters. This means that for some answers, one or more of the boxes will be empty.

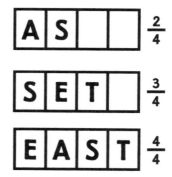

3. Encourage students to record new words, not those they already discovered in the first part of the activity. Again, have groups share the solutions for each word and record them on the class solutions list.

4. As a closing exercise, challenge students to find all of the fractional words in their names. Allow them to use their first, middle, or last names, and to rearrange the letters. If they are able, have them identify the fraction of their name that each word represents.

5. As a class, try to make two or three sentences using one fractional word from each student's name.

Connecting Learning

Part One

1. How many words did your group find that used half of the letters in the word *seat?* [1: at]
2. How many words did your group find that used three-fourths of the letters in the word *seat?* [2: sea, eat]
3. Were there any words that used less than half of the letters in the word *seat?* Explain. [yes, a]
4. What fractional words did you find in the other words you used?
5. Which word(s) had the most fractional words hidden within them? Why?

Part Two

1. What word did your group use? How many fractional words did you discover?
2. Do you think you have found them all? Why or why not?
3. Which kind of fractional words were most common—one half, one fourth, or three-fourths? Is this different from what you discovered in the first part of the activity? Why or why not?
4. Was it easier or harder to find fractional words when you could rearrange the letters? Why?
5. What fractional words were you able to find in your name? Is that more or less than most of your classmates?
6. Who was able to find the most words in his/her name? Why do you think this is?
7. Which kind of fractional words (one-half, two-thirds, etc.) were most common in the names of people in our class?

Extensions

1. Come up with a variety of writing challenges using the words discovered. For example, have students make a sentence using at least two of the fractional words found in *mate*, or write a short paragraph using all of the fractional words found in *seat*.
2. Challenge students to come up with their own words that can be broken down into fractional words, or to define all of the fractional words discovered for one word.
3. Use word chunks, such as *ate*, and have students try to find all of the words that can be made by adding one letter to the beginning of the chunk. [fate, late, rate, date, etc.]
4. Have students make a graph showing the frequency of each kind of fraction (one-fourth, one-half, two-thirds, three-fourths, etc.) in the words discovered.
5. Use words with five, seven, or nine letters to develop experience with uncommon fractions.
6. Give students the page on which they are challenged to match up a fractional word with the word it came from. Because there is more than one possible match for each word, students must look at the fraction listed by each fractional word to determine which match is correct. This extension uses fractions in their lowest terms, so students must be able to recognize $\frac{1}{3}$ as being equivalent to $\frac{2}{6}$, etc.

* Reprinted with permission from *Principles and Standards for School Mathematics*, 2000 by the National Council of Teachers of Mathematics. All rights reserved.

Key Questions

1. How many fractional words can you find within a large word?
2. How many different words can you make using the same set of letters, and what fraction of the letters do they use?

Learning Goals

Students will:

1. find all of the words hidden within a larger word that use only a fraction of the original letters—without rearranging the order, and

2. use the letters from the original words to create as many words as possible by arranging the letters in any order.

A	D	E	E	H
K	M	N	O	P
R	S	T	W	

A	D	E	E	H
K	M	N	O	P
R	S	T	W	

15

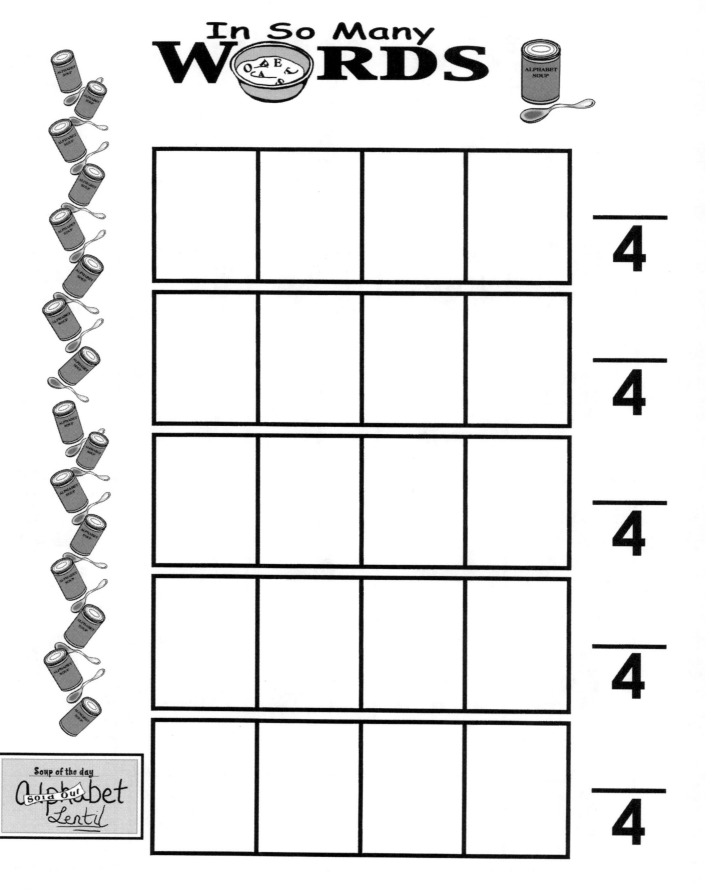

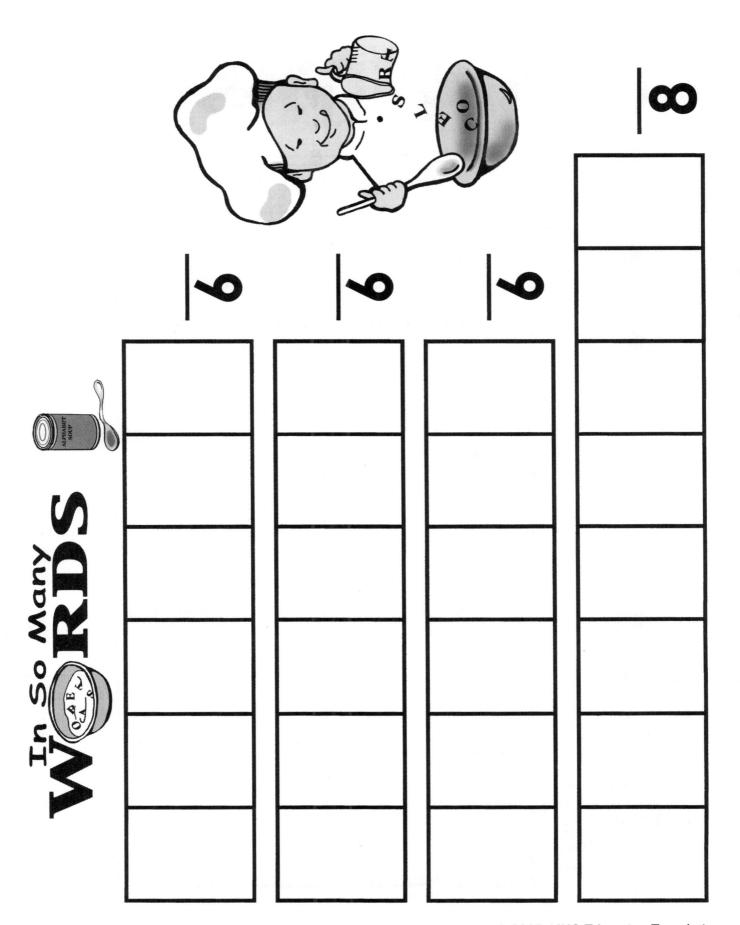

In So Many WORDS

Extension

Each word on the right is found inside two words on the left. Use the fractions to find the right match for all of the words. One example has been done for you.

1. Kite __f__

2. Late _____

3. Cat _____

4. Plants _____

5. Want _____

6. Send _____

7. Ending _____

8. Sit _____

9. Mitten _____

a. it ($\frac{1}{3}$)

b. ant ($\frac{1}{2}$)

c. end ($\frac{1}{2}$)

d. it ($\frac{2}{3}$)

e. at ($\frac{1}{2}$)

f. it ($\frac{1}{2}$)

g. ant ($\frac{3}{4}$)

h. at ($\frac{2}{3}$)

i. end ($\frac{3}{4}$)

Connecting Learning

Part One

1. How many words did your group find that used half of the letters in the word *seat?*

2. How many words did your group find that used three-fourths of the letters in the word *seat?*

3. Were there any words that used less than half of the letters in the word *seat?* Explain.

4. What fractional words did you find in the other words you used?

5. Which word(s) had the most fractional words hidden within them? Why?

Connecting Learning

Part Two

1. What word did your group use? How many fractional words did you discover?

2. Do you think you have found them all? Why or why not?

3. Which kind of fractional words were most common—one half, one fourth, or three-fourths? Is this different from what you discovered in the first part of the activity? Why or why not?

4. Was it easier or harder to find fractional words when you could rearrange the letters? Why?

5. What fractional words were you able to find in your name? Is that more or less than most of your classmates?

6. Who was able to find the most words in his/her name? Why do you think this is?

7. Which kind of fractional words (one-half, two-thirds, etc.) were most common in the names of people in our class?

FLIPPING OVER SYMMETRY

Topic
Symmetry

Key Questions
1. How many ways can you put two of the same pattern block together?
2. How many lines of symmetry are there in your solutions?

Learning Goals
Students will:
1. identify the shape in a set that is not congruent with the others,
2. recognize that shapes remain the same even when they are flipped and/or rotated,
3. discover all of the ways to put two of the same pattern block shape together, and
4. find all of the lines of symmetry that exist in the two-pattern block shapes they discovered.

Guiding Documents
Project 2061 Benchmark
- *Many objects can be described in terms of simple plane figures and solids. Shapes can be compared in terms of concepts such as parallel and perpendicular, congruence and similarity, and symmetry. Symmetry can be found by reflection, turns, or slides.*

NCTM Standards 2000*
- *Recognize and apply slides, flips, and turns*
- *Recognize and create shapes that have symmetry*
- *Identify and describe line and rotational symmetry in two- and three-dimensional shapes and designs*
- *Build new mathematical knowledge through problem solving*

Math
Geometry
  symmetry
  congruence
  transformations
    slides, flips, turns
Problem solving

Problem-Solving Strategy
Use manipulatives

Materials
Pattern blocks
Colored pencils or crayons
Materials for recording solutions (see *Management 3*)
Mirrors, one per group
Straight edges or rulers
Student pages

Background Information
The concepts of shape and symmetry are explored in this activity using the familiar pattern block manipulative. This activity is divided into two sections. The first has students explore transformations of congruent shapes, and the second has students explore line symmetry in simple shapes. This activity should not be students' first exposure to the concept of symmetry; they will need to have prior experience determining lines of symmetry in order to complete the second section of the activity.

Students should understand that a shape has a line of symmetry if it can be divided into two identical halves. This can be determined by using a mirror. If the image in the mirror is the same as the shape behind the mirror, then a line of symmetry exists.

Management
1. Students will work individually in *Part One*, and in small groups of three or four in *Part Two*. You will need enough pattern blocks for each student to have two of every shape except the tan rhombus.
2. The third student page is provided for students to record their solutions. Make mulitple copies of the page on card stock and cut the sections apart for students.
3. To record solutions, select one of the following methods:
   - Have students trace around the outlines of their pattern blocks to record each solution. These solutions can then be colored to correspond to the actual pieces.
   - If you have access to an Ellison machine, use the pattern block die cut to cut out the shapes, which students can then color and paste onto the solution papers. Alternatively, cut out each shape in the appropriate color.
   - Cut out sponges in the shapes of the pattern blocks (by hand, or using the Ellison machine), and have students sponge paint their solutions onto the papers using the corresponding colors.

Procedure

Part One

1. Distribute the first page; colored pencils; and two green triangle, blue rhombus, and red trapezoid pattern blocks to each student.

2. Explain the challenge, and define "congruent," if necessary (same size and shape). Encourage students to create the shapes using pattern blocks and rotate/flip them to help determine which one is not congruent. Discuss with students why a shape (such as the rhombus formed by connecting two triangles) is the same (congruent) even when it has been flipped and/or rotated. [The number of sides is the same, the length of the sides is the same, the angles are the same, etc.] If necessary, do a few more examples to solidify this concept in their minds.

3. Distribute the second student page to students and again encourage them to create each shape with pattern blocks and rotate and/or flip it to find its pair.

Part Two

1. Have students get into small groups and give each group a selection of pattern blocks that does not included the tan rhombuses.

2. Instruct students to take two triangles from the selection of pattern blocks. Ask them to put these two pieces together so that the edges line up completely.

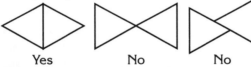

Yes   No   No

3. Challenge them to discover every way that two triangles can be put together following this rule. [Students should quickly realize that there is only one way that two triangles can be put together so that their edges line up.]

4. Repeat this process with squares and hexagons. Instruct students to leave their solutions assembled for later recording.

5. Have each group determine all the unique ways in which two rhombuses and two trapezoids can be put together. (There are multiple solutions.) Be sure students understand that when they are putting the trapezoid pieces together, they must line up a short edge with a short edge or a long edge with a long edge, never a short edge to a long edge. Again, each solution should be left intact for recording.

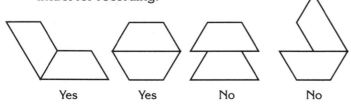

Yes   Yes   No   No

6. Invite groups to compare their solutions to see if everyone discovered the same ones. Determine if any discrepancies are actually new solutions, or merely flips and/or rotations of other solutions. Students should be able to discover two ways to combine the blue rhombus pieces and five ways to combine the trapezoids.

7. If any solutions are not discovered by the groups, lead the students in a time of class discovery to determine the remaining shapes.

8. Provide the solution-recording papers to each group and the necessary materials for students to record their solutions. (See *Management 3* for suggested methods.) Allow time for groups to record all of their solutions.

9. After all of the solutions have been recorded, distribute mirrors and straight edges to groups and review the concept of line (or mirror) symmetry. Go over the procedure for using a mirror to find lines of symmetry.

10. Have students work together in groups to discover lines of symmetry on each solution. Each line that is discovered should be drawn in with a ruler.

11. Compare the solutions and lines of symmetry from all of the groups and assist students in discovering any missing lines.

Connecting Learning

Part One

1. How could you tell which shape did not belong in the set (was not congruent)?

2. Did it help to have the pattern blocks to work with? Why or why not?

3. Was it difficult to find the congruent shapes on the second page? Why or why not?

Part Two

1. How many different ways did you find to put two triangles together? [Only one is possible.]

2. How about two squares? [one] …two hexagons? [one] …two rhombuses? [two] …two trapezoids? [five]

3. How many lines of symmetry were you able to find in each of your solutions? (See *Solutions*.)

4. How do you know you have found them all?

Extensions

1. Use more than one kind of pattern block to create shapes with symmetry.

2. Challenge students to make a shape that has two lines of symmetry or more.

3. Create real graphs using the individual solution papers. The graphs can be organized by shape, by number of lines of symmetry, by number of sides, etc.

Solutions

The solutions and lines of symmetry for each of the pattern block pieces are shown here.

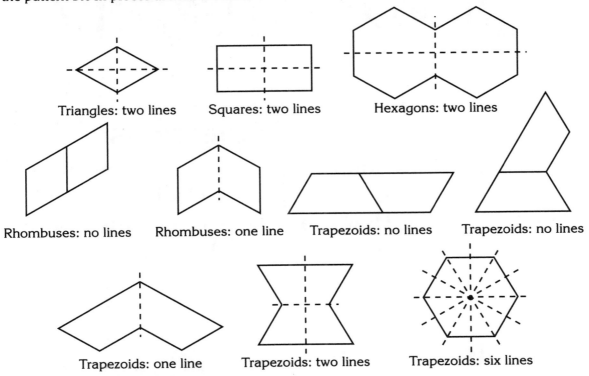

Triangles: two lines   Squares: two lines   Hexagons: two lines

Rhombuses: no lines   Rhombuses: one line   Trapezoids: no lines   Trapezoids: no lines

Trapezoids: one line   Trapezoids: two lines   Trapezoids: six lines

*   Reprinted with permission from *Principles and Standards for School Mathematics*, 2000 by the National Council of Teachers of Mathematics. All rights reserved.

FLIPPING OVER SYMMETRY

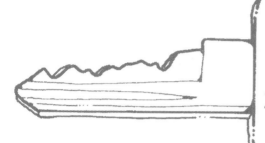

Key Questions

1. How many ways can you put two of the same pattern blocks together?
2. How many lines of symmetry are there in your solutions?

Learning Goals

Students will:

1. identify the shape in a set that is not congruent with the others,
2. recognize that shapes remain the same even when they are flipped and/or rotated,
3. discover all of the ways to put two of the same pattern block shape together, and
4. find all of the lines of symmetry that exist in the two-pattern block shapes they discovered.

24

FLIPPING OVER SYMMETRY

One shape in each set is not the same as the others.
Circle the shape that is different (not congruent).

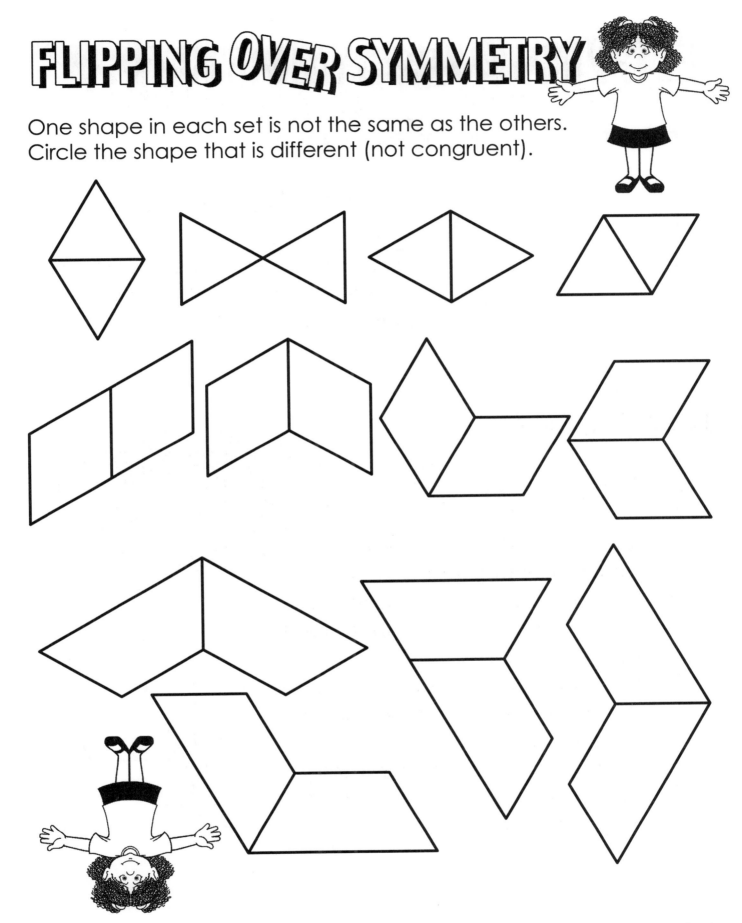

FLIPPING OVER SYMMETRY

Every shape here has a pair that is exactly the same (congruent). Find the pairs and color them the same color.

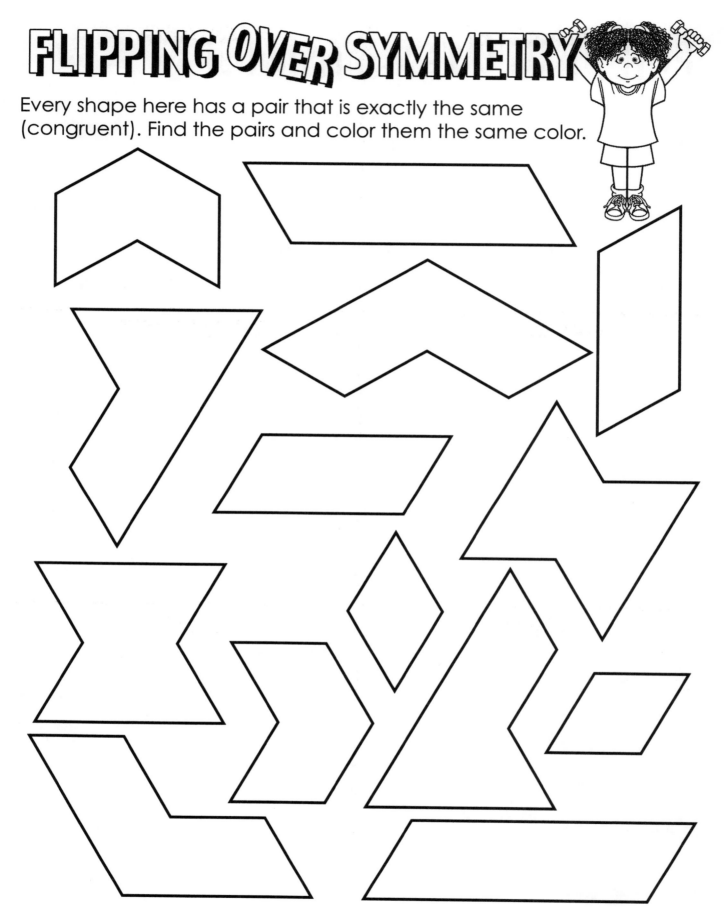

TRIANGLES

SQUARES

HEXAGONS

RHOMBUSES

TRAPEZOIDS

TRAPEZOIDS

FLIPPING OVER SYMMETRY

Connecting Learning

Part One

1. How could you tell which shape did not belong in the set (was not congruent)?

2. Did it help to have the pattern blocks to work with? Why or why not?

3. Was it difficult to find the congruent shapes on the second page? Why or why not?

Part Two

1. How many different ways did you find to put two triangles together?

2. How about two squares? ...two hexagons? ...two rhombuses? ...two trapezoids?

3. How many lines of symmetry were you able to find in each of your solutions?

4. How do you know you have found them all?

Problem-Solving Strategies

☐ + ☐ = ☐ **Write a Number Sentence**

Sometimes it helps to write down the parts of a problem as a number sentence. Then you can see if what you are doing makes sense. Are you using the right numbers? Are you adding where you should be subtracting? If you have a number sentence, you can answer these questions.

SOLVE IT! 3rd   29

Topic
Whole number operations

Key Question
What are all of the correct addition and subtraction problems that you can make with each set of numbers?

Learning Goals
Students will:
1. use sets of four numbers to make correct addition and subtraction problems,
2. begin to understand the commutative property of addition and subtraction, and
3. organize their solutions to determine if they have found them all.

Guiding Documents
Project 2061 Benchmark
- *Readily give the sums and differences of single-digit numbers in familiar contexts where the operation makes sense to them and they can judge the reasonableness of the answer.*

NCTM Standards 2000*
- *Understand various meanings of addition and subtraction of whole numbers and the relationship between the two operations*
- *Understand the effects of adding and subtracting whole numbers*
- *Illustrate general principles and properties of operations, such as commutativity, using specific numbers*
- *Develop fluency with basic number combinations for addition and subtraction*
- *Build new mathematical knowledge through problem solving*

Math
Number and operations
    addition
    subtraction
Commutative property
Problem solving

Integrated Processes
Observing
Recording
Comparing and contrasting
Organizing
Generalizing

Problem-Solving Strategies
Write a number sentence
Organize the information

Materials
Student page
Chart paper

Background Information
The activity consists of sets of four numbers from which students are to make addition and subtraction problems. Each set of four numbers can make several addition and subtraction problems. For example, one problem gives the numbers 5, 3, 2, and 8. Using these numbers it is possible to make the following problems: 2 + 3 = 5, 3 + 2 = 5, 3 + 5 = 8, 5 + 3 = 8, 5 – 3 = 2, 5 – 2 = 3, 8 – 5 = 3, and 8 – 3 = 5.

Not only does this problem help children become familiar with addition and subtraction fact families, its multiple solutions open the door for some meaningful discussions about the commutative property of addition and subtraction, framed in a way that children can understand.

Management
1. Each student will need his or her own copy of the student page.
2. It is recommended that you use chart paper to record the solutions. This will allow you to post them in the class or keep for future reference. You will need one page for the original list of solutions, and a second page for the organized list of solutions.

Procedure
1. Distribute the student pages and go over the instructions with the class.
2. Give students time to work on the problems and come up with two solutions for each set of numbers.
3. After students have completed the problems individually, ask them to share their answers with the class. On a piece of chart paper, record each unique solution discovered by students for all sets of numbers.

4. After a master list of solutions is compiled, move into a time of class discussion. If your list contains the same numbers used in both addition and subtraction problems, point this out to students and ask them how this can be. If none of the addition and subtraction problems uses the same numbers, encourage students to think about the addition problems they wrote down for each set of numbers and how those could be changed into subtraction problems.

5. As a class, try to determine if there are any solutions that are missing. Develop a way to organize the solutions to make this easier to determine.

6. Add any missing solutions to the reorganized list.

Connecting Learning

1. What addition problems did you come up with for the numbers 1, 8, 7, and 6? ...the numbers 3, 9, 6, and 3? ...the numbers 5, 3, 2, and 8?

2. Does the order of the numbers you are adding together matter? [No.] Why or why not? [2 + 3 is the same as 3 + 2. Both give you five, it doesn't matter in which order you add them.]

3. How do these problems compare to the subtraction problems you came up with? [Many of the addition and subtraction problems use the same numbers.]

4. Are there any number combinations that do not have both addition and subtraction problems?

5. Do you think these addition/subtraction problems could be changed into subtraction/addition problems? How would we do that?

6. Is it always possible to turn any addition problem into a subtraction problem? [Yes.] How? [Rearrange the numbers—5 + 3 = 8, 8 – 5 = 3 or 8 – 3 = 5.]

7. What do you have to do to change a subtraction problem into an addition problem? [Rearrange the numbers—7 – 2 = 5, 5 + 2 = 7] Does this always work? [Yes.]

8. Do you think we have found all of the possible solutions? Why or why not?

9. How could we organize our list of solutions to make it easier to determine if we are missing any?

Extensions

1. Ask students to make their own problems by choosing four numbers that can all be used in some combination. Challenge them to trade problems with classmates and find every possible addition and subtraction problem that can be made using those numbers.

2. Give students five or six numbers from which to create their problems instead of only four.

3. Challenge students to create problems where more than two numbers are added or subtracted (1 + 2 + 3 = 6).

4. Have students use four numbers to create problems using both addition and subtraction (8 + 4 – 5 = 7).

* Reprinted with permission from *Principles and Standards for School Mathematics*, 2000 by the National Council of Teachers of Mathematics. All rights reserved.

I+ A ‖
Adds Up

Key Questions

What are all of the correct addition and subtraction problems that you can make with each set of numbers?

Learning Goals

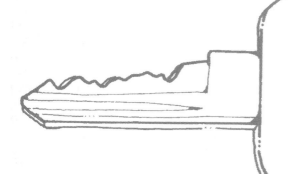

Students will:

1. use sets of four numbers to make correct addition and subtraction problems,
2. begin to understand the commutative property of addition and subtraction, and
3. organize their solutions to determine if they have found them all.

I+ A = Adds Up

Use the numbers to write two correct math problems.

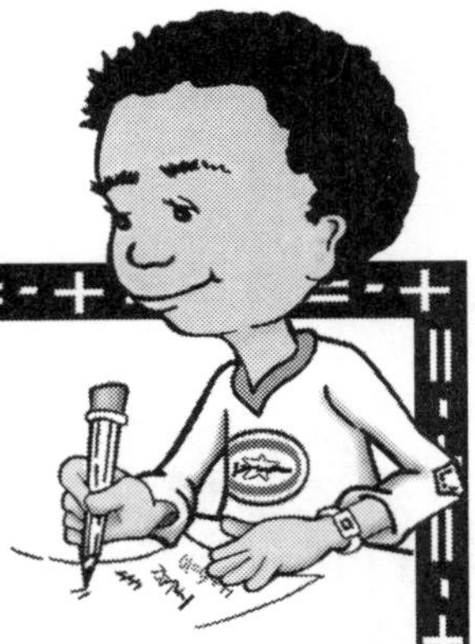

1 8 7 6

_____ + _____ = _____

_____ – _____ = _____

3 9 6 3

_____ + _____ = _____

_____ – _____ = _____

5 3 2 8

_____ + _____ = _____

_____ – _____ = _____

4 7 1 3

_____ + _____ = _____

_____ – _____ = _____

6 1 4 5

_____ + _____ = _____

_____ – _____ = _____

2 9 5 7

_____ + _____ = _____

_____ – _____ = _____

9 1 7 8

_____ + _____ = _____

_____ – _____ = _____

2 4 6 8

_____ + _____ = _____

_____ – _____ = _____

Connecting Learning

1. What addition problems did you come up with for the numbers 1, 8, 7, and 6? …the numbers 3, 9, 6, and 3? …the numbers 5, 3, 2, and 8?

2. Does the order of the numbers you are adding together matter? Why or why not?

3. How do these problems compare to the subtraction problems you came up with?

4. Are there any number combinations that do not have both addition and subtraction problems?

5. Do you think these addition/ subtraction problems could be changed into subtraction/addition problems? How would we do that?

Connecting Learning

6. Is it always possible to turn any addition problem into a subtraction problem? How?

7. What do you have to do to change a subtraction problem into an addition problem? Does this always work?

8. Do you think we have found all of the possible solutions? Why or why not?

9. How could we organize our list of solutions to make it easier to determine if we are missing any?

THAT'S SUM NAME!

Topic
Problem solving

Key Question
Using the alphabet chart provided, how many points is your name worth?

Learning Goal
Students will write number sentences to calculate the sums of the values of the letters in various names.

Guiding Document
NCTM Standards 2000*
- *Develop fluency in adding, subtracting, multiplying, and dividing whole numbers*
- *Build new mathematical knowledge through problem solving*
- *Solve problems that arise in mathematics and in other contexts*
- *Apply and adapt a variety of appropriate strategies to solve problems*
- *Monitor and reflect on the process of mathematical problem solving*

Math
Number and operations
    addition
Problem solving

Integrated Processes
Observing
Collecting and recording data
Comparing and contrasting

Problem-Solving Strategy
Write a number sentence

Materials
Student pages
Overhead of student page, optional

Background Information
The letters in the English language are not all used equally in speech and writing. The most commonly used letters, in order of their frequency, are e, t, a, o, i, n, s, h, r, d, l, and u. The least commonly used letters are w, b, v, k, x, j, q, and z. In this activity, the letters have been given point values from 1 to 26 from the most frequently used to the least.

Management
1. An overhead transparency of the student page can be used to introduce this activity to the class.
2. This activity can be done individually or with students working in groups.

Procedure
1. Distribute the student pages.
2. Introduce the activity and monitor students as they work.
3. Facilitate a whole class sharing session at the end of the activity.

Connecting Learning
1. How many points is your first name worth?
2. How many points is your last name worth?
3. How many points is your whole name worth?
4. What did you do to find the value of your name? [I wrote a number sentence.]
5. How did you find the name with the greatest point value? [I tried to use higher point letters, I looked at the names I had done and picked the highest value, etc.]
6. How did you find the name with the point value closest to 100?

Extensions
1. Give students target numbers, like 100 points or 50 points, and have them try to find words with letter values that add to these sums.
2. Have students try to write an 8-10-word sentence with a high point value.
3. Have students try to write a sentence of at least 10 words with a low point value.
4. Have students cut out the letters and their values on this sheet and use them as letter tiles for the game of Scrabble®.
5. Have students order their names from least to greatest point value.

Solutions
Some 50-point names include *Zach* and *Shakita*. *Orlando Bloom* is worth exactly 100 points.

Home Link
Have students take the point chart home and work with their families to find the point values of families' or friends' names.

* Reprinted with permission from *Principles and Standards for School Mathematics*, 2000 by the National Council of Teachers of Mathematics. All rights reserved.

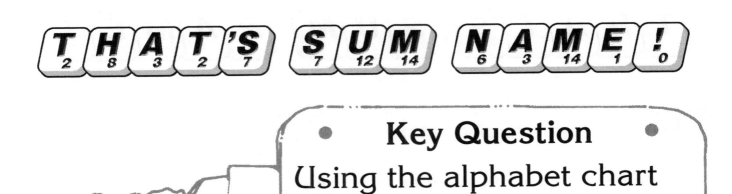
Key Question

Using the alphabet chart provided, how many points is your name worth?

Learning Goal

Students will:

write number sentences to calculate the sums of the values of the letters in various names.

THAT'S SUM NAME!

How many points is your name worth? The following table lists the letters in the alphabet according to how often they are used in written English. Each letter is given a point value from 1 to 26 based on its frequency.

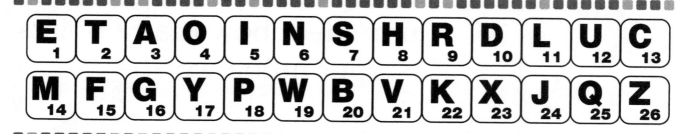

E	T	A	O	I	N	S	H	R	D	L	U	C
1	2	3	4	5	6	7	8	9	10	11	12	13
M	F	G	Y	P	W	B	V	K	X	J	Q	Z
14	15	16	17	18	19	20	21	22	23	24	25	26

Using this table, find the sums of the letter values in your first name, last name, and your whole name. (Problem-solving hint: Write a number sentence.)

Think of some of your favorite super heroes, cartoon character, sports stars, singers, etc. Write their names below and then find out how many points their names are worth.

Challenge One:

Find or make up a name with as great a point value as you can.

Challenge Two:

Find or make up a name with a point value as close to 50 as you can.

Super challenge:

Find a name that is worth 100 points.

THAT'S SUM NAME!

Connecting Learning

1. How many points is your first name worth?

2. How many points is your last name worth?

3. How many points is your whole name worth?

4. What did you do to find the value of your name?

5. How did you find the name with the greatest point value?

6. How did you find the name with the point value closest to 100?

What is the One?

Topic
Fractional parts and relationships

Key Question
Why do you need to know what the one is when working with fractions?

Learning Goals
Students will:
1. identify relationships in fractions, and
2. explain how relationships change when the one changes.

Guiding Document
NCTM Standards 2000*
- *Develop understanding of fractions as parts of unit wholes, as parts of a collection, as locations on number lines, and as divisions of whole numbers*
- *Use models, benchmarks, and equivalent forms to judge the size of fractions*
- *Build new mathematical knowledge through problem solving*

Math
Fractions
Problem solving

Integrated Processes
Observing
Comparing and contrasting
Communicating
Recording data

Problem-Solving Strategies
Write a number sentence
Use manipulatives

Materials
Pattern blocks (see *Management 1*)
What is the One rubber band book (see *Management 3*)

Background Information
The identification of the *one* or the unit whole is a big idea in the study of fractions and fractional relationships. In order for the students to be able to identify fractional parts, they need to be able to identify the whole or the one. In this experience, students will be exploring fractional relationships using pattern blocks. The students will use the hexagon, the trapezoid, the blue rhombus, and the equilateral triangle from the traditional pattern block set. If the AIMS fractional pattern block pieces are available, the students can use these to explore other fractional relationships.

Management
1. For each student group, you will need three hexagons, two trapezoids, three blue rhombuses, and six equilateral triangles.
2. Student groups of two work well for this activity.
3. A rubber band book is constructed by folding the student pages in half horizontally and vertically, nesting them so the pages go in order, and holding them together with a number 19 rubber band.

Procedure
1. Ask the *Key Question* and state the *Learning Goals*.
2. Place a hexagon pattern block piece on the surface of the overhead projector and tell the students that the value of the piece is one. Place a trapezoid on the surface of the projector, and ask the students what the value of the trapezoid would be if the hexagon is the one. (The students should tell you that the trapezoid is one-half if the hexagon is the one, or the whole, since it would take two identical trapezoids to make one hexagon.)
3. Place other pattern block pieces on the overhead and have the students identify their values if the hexagon is the one.
4. Place the trapezoid on the overhead and tell the students that this is now the one. Place the hexagon back on the overhead and ask them what the value of the hexagon would be. [The students should be able to state the value of the hexagon is now two.]
5. Distribute the rubber band book *What is the One?* Direct the students to record relationships about the one listed on each page.

Connecting Learning
1. Why is it important to know what the one is?
2. How can a hexagon have different values?
3. What other "ones" do we use in our lives? [a cup, a foot or a yard, other units of measurement, etc.]
4. How did recording your observations help you in this activity?
5. How would you describe how the "one" is related to fractions?

* Reprinted with permission from *Principles and Standards for School Mathematics*, 2000 by the National Council of Teachers of Mathematics. All rights reserved.

What is the One?

Key Questions

Why do you need to know what the one is when working with fractions?

Learning Goals

Students will:

1. identify relationships in fractions, and
2. explain how relationships change when the one changes.

What is the One?

Tell what you learned about fractions by doing this activity.

If this is the ONE, what other relationships can you describe?

You name the ONE and tell about other relationships.

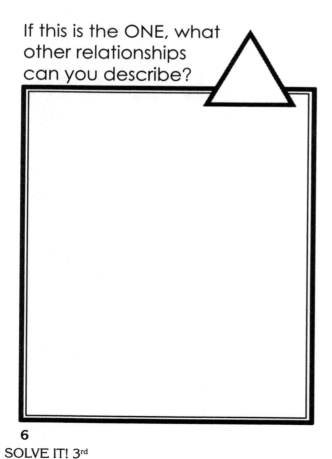

If this is the ONE, what
other relationships
can you describe?

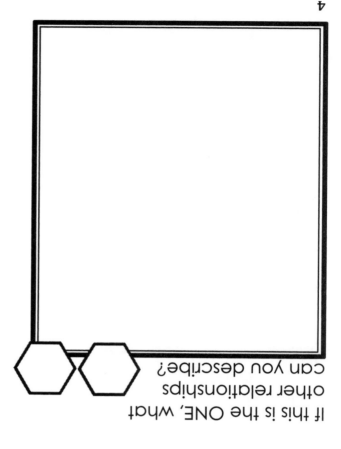

If this is the ONE, what
other relationships
can you describe?

If this is the ONE, what
other relationships
can you describe?

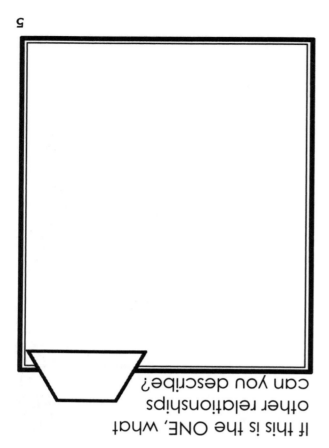

If this is the ONE, what
other relationships
can you describe?

Connecting Learning

1. Why is it important to know what the one is?

2. How can a hexagon have different values?

3. What other "ones" do we use in our lives?

4. How did recording your observations help you in this activity?

5. How would you describe how the "one" is related to fractions?

Problem-Solving Strategies
Draw out the Problem

Drawing pictures is a useful problem-solving tool. Pictures help you keep track of important information. They also help when a problem has lots of details. With a picture you can see every part of the problem at once. Keep the pictures simple. You don't need to spend lots of time drawing.

Schmoos 'n' Goos

Topic
Problem solving

Key Question
How many schmoos and goos are there for the given number of bodies and legs?

Learning Goal
Students will use the problem-solving strategies of using manipulatives and drawing pictures to solve several story problems.

Guiding Document
NCTM Standards 2000*
- *Build new mathematical knowledge through problem solving*
- *Solve problems that arise in mathematics and in other contexts*
- *Apply and adapt a variety of appropriate strategies to solve problems*
- *Monitor and reflect on the process of mathematical problem solving*

Math
Problem solving

Integrated Processes
Observing
Classifying
Collecting and recording data
Generalizing

Problem-Solving Strategies
Draw out the problem
Use manipulatives

Materials
Toothpicks, 16 per student
Student pages

Background Information
The story problems posed here are like the ones algebra students encounter. In algebra, these problems are solved using two equations, each with two unknowns. However, elementary school students can solve these problems—if they use the right problem-solving strategies. The two strategies highlighted in this activity are using manipulatives and drawing pictures. To illustrate these strategies, consider the story problem in *Part One*. This problem says that there are animals called schmoos and goos. The only difference between these animals is that the schmoos have four legs and the goos have two. The problem asks: If there are five animals and 16 legs, how many of each animal is there?

To solve this problem algebraically, you set up two simultaneous equations. If the schmoos are represented by the letter s and goos are represented by the letter g, the two equations are $s + g = 5$ (the equation for the number of bodies) and $4s + 2g = 16$ (the equation for the number of legs). Solving these equations simultaneously shows that $s = 3$ and $g = 2$.

Elementary students can solve this same problem using 16 toothpicks and the pictures shown on the student page. To do this, the students start by placing the toothpick "legs" underneath the five pictures two at a time using 10 toothpicks. They then add two more toothpicks to three of the drawings. This shows that there are three four-legged schmoos, and two two-legged goos. In *Part Two*, the students use what they learned in *Part One* to solve similar problems. This time, however, they draw circles to represent the animals' bodies and draw lines to represent their legs.

Management
1. This activity can be done individually or in groups.
2. If this is your students' first exposure to this type of problem, you may want to do *Part One* together as a class before having students do *Part Two*.

Procedure
Part One
1. Distribute the first student page and toothpicks.
2. Make sure students understand the problem and then facilitate their problem-solving efforts.
3. When students have solved the problem, have a class discussion to allow them to share their mathematical thinking.

Part Two
1. Distribute the second student page and make sure students understand the problems.
2. Explain that for these problems, they will use the same mathematical thinking they used in *Part One*, but that they will draw circles and lines to represent the animals' bodies and legs, respectively.
3. Facilitate students' problem-solving efforts as they work on the problems.
4. When students have finished with the story problems presented on the page, have them make up their own problems on the back of the paper. Students can then solve these problems themselves or exchange them with someone else.

Connecting Learning

Part One
1. How many schmoos are there? [3]
2. How many goos are there? [2]
3. How did you solve this problem?

Part Two
1. How many schmoos and goos in question one? [1 schmoo and 3 goos]
2. How many schmoos and goos in question two? [2 schmoos and 6 goos]
3. How many schmoos and goos in question three? [7 schmoos and 0 goos]
4. How many schmoos and goos in question four? [7 schmoos and 2 goos]
5. What strategies did you use when solving these problems?
6. What do you notice about the number of legs in each problem? [The number of legs is always even.]
7. Why is there always an even number of legs? [Although you can have an odd number of animals, the legs come in sets of twos and fours. When an odd number is multiplied by an even number, the product (number of legs) is still an even number.]

Extensions
1. Have students make tables for all the possible combinations of schmoos and goos for a given number of bodies.
2. Help students set up number sentences to represent the problems.

---

* Reprinted with permission from *Principles and Standards for School Mathematics*, 2000 by the National Council of Teachers of Mathematics. All rights reserved.

Schmoos 'n' Goos

Learning Goal

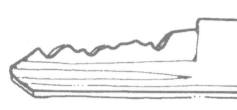

Students will:

use the problem-solving strategies of using manipulatives and drawing pictures to solve several story problems.

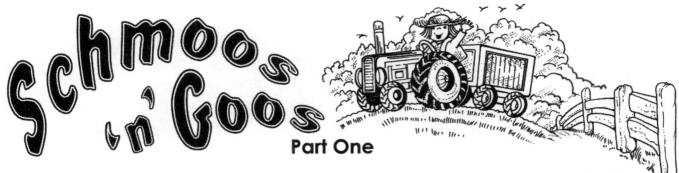

Part One

While visiting the Schmoogoole Ranch, you see 5 animals behind a fence. Rancher Rachel reports that the animals are schmoos and goos. These animals look the same except that the schmoos have 4 legs and goos have 2. She tells you there are 16 legs behind the fence. Can you find how many schmoos and goos there are?

Use toothpicks to add "legs" to the picture below to find out how many schmoos and goos are behind the fence.

How many schmoos are there?   How many goos are there?

How did you solve this problem?

Part Two

Use what you learned in *Part One* to answer the following questions. Remember that schmoos have 4 legs and goos have 2.

Problem-solving hint: Draw simple pictures to help. Use circles for the schmoos' and goos' bodies and lines for their legs.

1. There are 4 bodies and 10 legs. How many schmoos and goos are there?

2. There are 8 bodies and 20 legs. How many schmoos and goos are there?

3. There are 7 bodies and 28 legs. How many schmoos and goos are there?

4. There are 9 bodies and 32 legs. How many schmoos and goos are there?

Challenge: Using schmoos and goos, make up you own problems on the back of this paper. You can solve them yourself or give them to someone else to solve.

Connecting Learning

Part One

   1. How many schmoos are there?

   2. How many goos are there?

   3. How did you solve this problem?

Part Two

1. How many schmoos and goos in question one?

2. How many schmoos and goos in question two?

3. How many schmoos and goos in question three?

4. How many schmoos and goos in question four?

5. What strategies did you use when solving these problems?

6. What do you notice about the number of legs in each problem?

7. Why is there always an even number of legs?

SAWING LOGS

Topic
Problem solving

Key Questions
1. How many cuts does it take to saw a log into five equal lengths?
2. How can you use this information to help you solve more difficult problems?

Learning Goals
Students will:
1. determine how many cuts are needed to saw a log into various equal lengths,
2. use this information to help them solve a more difficult problem, and
3. see the importance of drawing pictures when solving certain types of problems.

Guiding Document
NCTM Standards 2000*
- *Build new mathematical knowledge through problem solving*
- *Solve problems that arise in mathematics and in other contexts*
- *Apply and adapt a variety of appropriate strategies to solve problems*

Math
Problem solving

Integrated Processes
Observing
Recording
Generalizing

Problem-Solving Strategies
Draw out the problem
Wish for an easier problem

Materials
Student page

Background Information
The culminating question in this activity is this: *A farmer needs to build a fence 100 meters long. In order to support the fence, there must be one post every five meters. How many posts does she need to build the fence?* Before you read on, *please* work on this problem right now. How did you come up with your solution? If you're like many people, you knew that you could simply divide the 100 meters by the five meters to find that 20 posts are necessary. Was this your answer? If so, you join the majority of the people who get this problem wrong. If you didn't get 20 and know you have the correct answer, good for you! You didn't fall into the trap of blindly applying an arithmetic operation—in this case division—to solve the problem.

Let's try a simpler problem (an important problem-solving skill) to see where most people go wrong on this problem. What if the fence in the problem was only 10 meters instead of 100? Draw a picture of this fence. How many posts do you need? As soon as you draw the picture you realize that three posts are needed and that you can't simply divide 10 by five to get the answer since you must have a post to anchor the starting (zero) point of the fence. Drawing a picture is the key problem-solving strategy needed to get the correct answer to this problem. If you got the above problem wrong the first time, you now see the answer is 21 posts.

This activity is designed to help elementary students see the importance of drawing pictures when solving certain types of problems. Before students are given the above problem, they are asked to solve several simpler, related problems—another important problem-solving strategy. In these initial problems, they are asked how many cuts are needed to saw a log into various equal lengths. Like the fence problem above, the answers to these initial problems can't be obtained by rotely applying arithmetic. Yet, as soon as a picture is drawn, the answer becomes obvious and simple arithmetic can then be applied.

After doing these introductory problems, students are given a table to fill in showing the numbers of lengths produced by various numbers of cuts. As they complete this table, they should realize that the number of lengths is one more than the number of cuts. This generalization, which can be written $n + 1$, should be recorded in the last column of the table under n cuts. If students have not been exposed to algebraic variables, be sure to use this opportunity to introduce this key concept here.

Management

1. This activity is best done in collaborative groups with plenty of interaction between students. In this way, students learn from each other as they practice mathematical communication and problem solving.
2. Each student will need his or her own copy of the student page.

Procedure

1. Distribute the student page and have students get into groups.
2. Provide enough time for groups to complete the questions.
3. Close with a time of whole-class discussion so that students can see the importance of employing problem-solving skills like drawing pictures and doing simpler, related problems.

Connecting Learning

1. Was your guess for the first question correct? Why or why not?
2. How many cuts does it take to divide a log into five equal pieces? [four] ...three equal pieces? [two] ...eight equal pieces? [seven]
3. How does the number of cuts relate to the number of pieces? [There is always one more piece than the number of cuts.]
4. How many fence beams does the farmer need to support her fence? [21]
5. What strategies did you use to solve this problem?
6. How is this problem like the log-sawing problems? How is it different?

* Reprinted with permission from *Principles and Standards for School Mathematics*, 2000 by the National Council of Teachers of Mathematics. All rights reserved.

SAWING LOGS

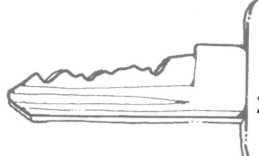

Key Questions

1. How many cuts does it take to saw a log into five equal lengths?
2. How can you use this information to help you solve more difficult problems?

Learning Goals

Students will:

1. determine how many cuts are needed to saw a log into various equal lengths,
2. use this information to help them solve a more difficult problem, and
3. see the importance of drawing pictures when solving certain types of problems.

SAWING LOGS

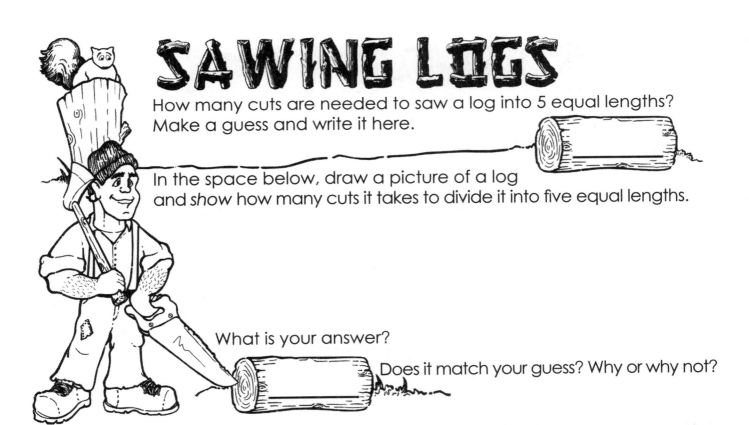

How many cuts are needed to saw a log into 5 equal lengths? Make a guess and write it here.

In the space below, draw a picture of a log and *show* how many cuts it takes to divide it into five equal lengths.

What is your answer?

Does it match your guess? Why or why not?

Using what you just learned, answer the following questions.

How many cuts would it take to saw a log into three equal lengths?

How about eight equal lengths?

Fill in the table below showing the number of lengths produced by different numbers of cuts. (The *n* in the table stands for *any* number of cuts.)

Cuts	1	2	3	4	5	6	7	8	n
Lengths									

Challenge:

Use what you learned on this page to solve the following problem. Show your work on the back of this page.

A farmer needs to build a fence 100 meters long. In order to support the fence, there must be one post every five meters. How many posts does she need to build the fence?

SAWING LOGS

Connecting Learning

1. Was your guess for the first question correct? Why or why not?

2. How many cuts does it take to divide a log into five equal pieces? ...three equal pieces? ...eight equal pieces?

3. How does the number of cuts relate to the number of pieces?

4. How many fence beams does the farmer need to support her fence?

5. What strategies did you use to solve this problem?

6. How is this problem like the log-sawing problems? How is it different?

Picturing Clues

Topic
Problem solving

Key Question
How can you solve word problems by drawing pictures?

Learning Goal
Students will solve word problems by drawing pictures.

Guiding Document
NCTM Standards 2000*
- *Apply and adapt a variety of appropriate strategies to solve problems*
- *Build new mathematical knowledge through problem solving*

Math
Problem solving

Integrated Processes
Observing
Comparing and contrasting
Recording data
Interpreting data
Drawing conclusions

Problem-Solving Strategy
Draw out the problem

Materials
Student pages
Transparencies of student pages
Colored pencils or crayons

Background Information
 A powerful problem-solving tool for younger learners is drawing out the problem. This strategy allows students to represent all of the necessary information, and gives them a concrete way to justify their responses. This activity simply presents several word problems that can be solved most easily by drawing out the problem.

Management
1. This activity is designed to be done whenever you have a few minutes for students to work individually.

2. Photocopy one set of the student pages onto transparencies and cut them apart for your own use. You may wish to draw in your own picture solutions ahead of time, or you can do this while students complete their pictures.
3. Copy and cut apart the student pages before doing this activity. Each student will need his or her own copy of the clue(s) being worked on at any given time.

Procedure
1. Select a problem to present to the class. Put your copy of this problem up on the overhead projector. (If you have already drawn the picture, be sure that it is covered with something so that students cannot see it.)
2. Distribute colored pencils or crayons and a copy of the problem to each student.
3. Give students time to draw pictures of their solutions.
4. Have a time of sharing where you reveal your picture and compare it to students' pictures.
5. If desired, post some of the pictures to show the different ways that students chose to illustrate the solutions.
6. Repeat this process with the additional problems as desired.

Connecting Learning
1. How did drawing out the picture help you answer the problems?
2. What are some other ways that you could have solved these problems?
3. Do you think that these other ways would be harder or easier than drawing the problem? Why?

Extensions
1. As students become comfortable with the process, they can be challenged to develop their own problems for classmates to solve.
2. Challenge students to use manipulatives to represent the parts of the problem rather than drawing each one out.

* Reprinted with permission from *Principles and Standards for School Mathematics*, 2000 by the National Council of Teachers of Mathematics. All rights reserved.

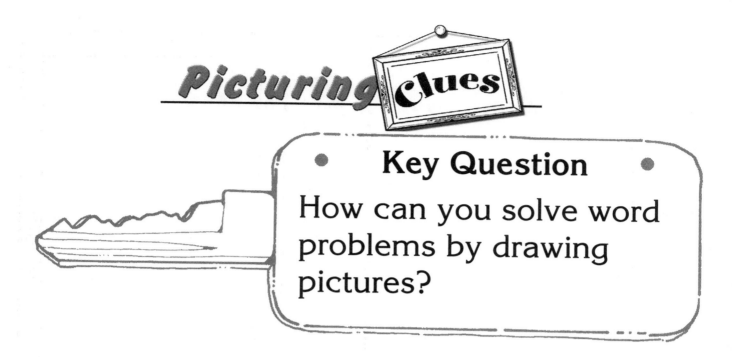

Picturing Clues

Key Question

How can you solve word problems by drawing pictures?

Learning Goal

Students will:

solve word problems by drawing pictures.

The number of wheels on three blue cars, four red tricycles, and a bright yellow bicycle. _____

The number of fingers and spoons at a table where two people are eating soup. _____

The number of antlers on three reindeer plus the number of tails on four kangaroos. _____

The number of legs on three dining room chairs and the people sitting in those chairs. _____

The number of sides on two triangles, a square, and two hexagons. _____

The number of even numbers on five dice. _____

The number of toes on four bouncing baby boys. _____

The number of sweets in a half dozen chocolate chip cookies and two dozen sugar donuts. _____

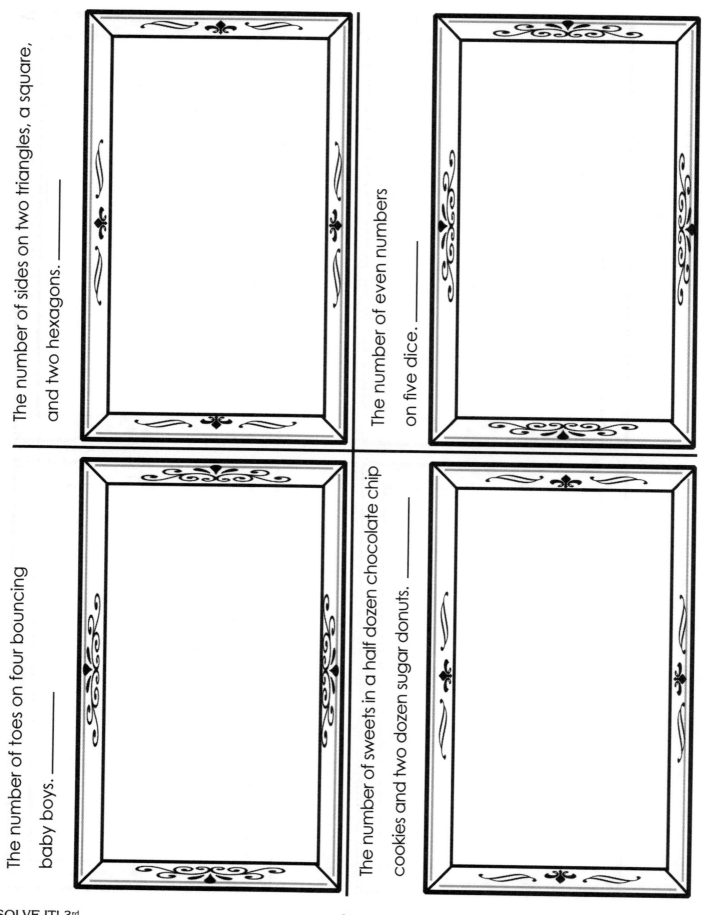

The number of ears on three little boys, two rabbits, and four ears of corn. _____

The number of faces on four quarters and two tired bankers. _____

The number of wings on three big white airplanes and five red cardinals. _____

The number of legs and tails on two little girls and three white cats. _____

Connecting Learning

1. How did drawing out the picture help you answer the problems?

2. What are some other ways that you could have solved these problems?

3. Do you think that these other ways would be harder or easier than drawing the problem? Why?

The Lily Pad Hop

Topic
Problem solving

Key Question
On what lily pads would both frogs land?

Learning Goal
Students will solve word problems by drawing pictures.

Guiding Document
NCTM Standards 2000*
- *Apply and adapt a variety of appropriate strategies to solve problems*
- *Build new mathematical knowledge through problem solving*

Math
Problem solving

Integrated Processes
Observing
Comparing and contrasting
Recording data
Interpreting data

Problem-Solving Strategy
Draw out the problem

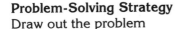

Materials
Problem card (see *Management 3*)
Paper for drawing
Colored pencils or crayons, optional

Background Information
Drawing out the problem is often a good way to solve word problems. This strategy allows students to see all parts of the problem and gives them a concrete way to justify their responses. Many students will begin by making detailed illustrations before realizing that quick sketches can deliver the message in a more timely fashion.

Management
1. This activity can be done individually or in pairs.
2. At the end of this activity, students are to share their solutions and analyze differences.

3. The student page has two identical problem cards. Each student or group will need one problem card.

Procedure
1. Distribute one problem card and paper for drawing to each student or group.
2. Give individuals or groups time to draw out the problem and determine solutions.
3. Allow students to share the answers they got and how they got those answers.

Connecting Learning
1. How did you figure out the problem?
2. How does drawing the problem out help you find the answer?
3. How was your drawing like other peoples'? How was it different?
4. On what lily pads did both frogs land?
5. How would numbering your lily pads make it easier to describe your answer?
6. Did everyone find the same answers? Explain. [Probably not. The numbers students get are determined by the side of the pond on which each frog begins. If Linus starts on the side closest to the number one lily pad, both frogs will land on the fourth, 10th, 16th, and 22nd lily pads. If Linus starts on the side closest to the number 24 lily pad, both frogs will land on the third, ninth, 15th, and 21st lily pads.]
7. How many leaps did each frog have to make? [Linus jumped on 12 pads and needed one more jump to get to the pond's bank. Lillian jumped on eight lily pads and needed one more to get to the pond's bank.]
8. What patterns can you find in the numbers of the lily pads on which both frogs landed? [One set is all odd numbers (3, 9, 15, 21); the other set is all even numbers (4, 10, 16, 22); there is a difference of six between the lily pads on which they landed.]

* Reprinted with permission from *Principles and Standards for School Mathematics*, 2000 by the National Council of Teachers of Mathematics. All rights reserved.

The Lily Pad Hop

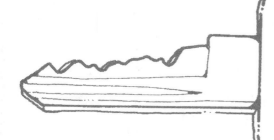

Key Question

On what lily pads would both frogs land?

Learning Goal

Students will:

solve word problems by drawing pictures.

65

Linus and Lillian were on opposite sides of the pond. Both wanted to switch sides. They decided to hop on the 24 lily pads that were on the pond.

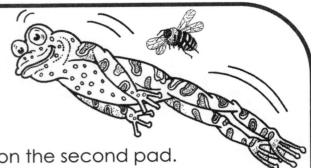

Linus can hop over one pad and land on the second pad. You could say he leaps by twos. Lillian, who is older, can hop over two pads and land on the third. She leaps by threes.

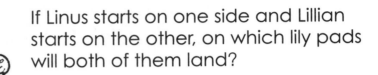

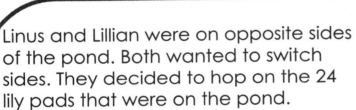

If Linus starts on one side and Lillian starts on the other, on which lily pads will both of them land?

Show how you can solve this problem.

Linus and Lillian were on opposite sides of the pond. Both wanted to switch sides. They decided to hop on the 24 lily pads that were on the pond.

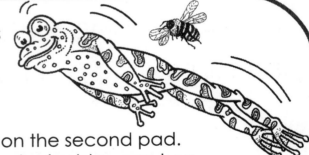

Linus can hop over one pad and land on the second pad. You could say he leaps by twos. Lillian, who is older, can hop over two pads and land on the third. She leaps by threes.

If Linus starts on one side and Lillian starts on the other, on which lily pads will both of them land?

Show how you can solve this problem.

The Lily Pad Hop

Connecting Learning

1. How did you figure out the problem?

2. How does drawing the problem out help you find the answer?

3. How was your drawing like other peoples'? How was it different?

4. On what lily pads did both frogs land?

5. How would numbering your lily pads make it easier to describe your answer?

6. Did everyone find the same answers? Explain.

7. How many leaps did each frog have to make?

8. What patterns can you find in the numbers of the lily pads on which both frogs landed?

Problem-Solving Strategies
Guess and Check

Sometimes to solve a problem, it's a good idea to just make a guess. Then you can check your answer to see if it's correct. If it's not, make another guess using what you learned from your first guess. Soon you will find the correct answer. This is a good strategy to use when you don't know how to approach a problem. It's also good when the problem is very complicated or has lots of answers.

Topic
Problem solving

Key Question
How many pumpkins can be enclosed with fences of different lengths?

Learning Goals
Students will:
1. use chains of paper clips (fences) to enclose patches of pumpkins,
2. discover the different numbers of pumpkins that can be enclosed with a fence of a given length,
3. describe the shape of each enclosure,
4. predict how many pumpkins they will be able to enclose with a larger fence, and
5. test their predictions.

Guiding Document
NCTM Standards 2000
- *Explore what happens to measurements of a two-dimensional shape such as its perimeter and area when the shape is changed in some way*
- *Build new mathematical knowledge through problem solving*
- *Solve problems that arise in mathematics and in other contexts*

Math
Perimeter
Area
Problem solving

Integrated Processes
Observing
Collecting and recording data
Comparing and contrasting
Predicting

Problem-Solving Strategies
Guess and check
Use manipulatives

Materials
Small paper clips, 16 per student
Transparency film
Student pages

Background Information
This activity indirectly explores perimeter and area in a problem-solving setting. Students are given chains of paper clips (fences) with which they are to enclose patches of pumpkins. They are challenged to discover the different numbers of pumpkins that can be enclosed with a fence of a given length, and to describe the shape of each enclosure. They are then challenged to predict how many pumpkins they will be able to enclose with a larger fence and test their predictions. This can lead to an understanding of the generalization that the largest area for a given perimeter is the one that is closest to a square (when the shapes are restricted to rectangles or squares). Students will also be able to recognize that as the perimeter increases, so does the largest possible area.

Management
1. Make a copy of the pumpkin patch page on transparency film for demonstration purposes.
2. Each student will need one copy of every student page and 16 small paper clips.

Procedure
1. Distribute 16 paper clips, the first student page, and the pumpkin patch page to each student.
2. Have students take 12 of their paper clips and link them together to form a closed loop. (Be sure that none of the paper clips are linked through the smaller interior hook.) Inform students that this loop is a fence with which they are going to enclose pumpkins in their pumpkin patches.
3. Use the overhead projector to model the proper way to enclose a patch of pumpkins. Each paper clip must follow the edge of a square surrounding a pumpkin. Paper clips must be in line with or at right angles to each other and form rectangles or squares; no circles, triangles, or other shapes that would cause partial pumpkins to be enclosed are allowed.
4. Have students find and record all the different-sized patches they can enclose with a 12-link fence. (They should be able to enclose five, eight, and nine pumpkins, depending on the shape of the fence.)

5. Once all solutions have been recorded, tell students that they are going to be adding two more links (paper clips) to their fences. Distribute the second student page. Ask them to predict how many pumpkins they will be able to enclose with these longer fences and record their predictions.

6. Repeat this process once more with a 16-link fence.

7. Once students have completed the exploration and recording time, conduct a time of class discussion where students can generalize some of their findings. Students should be able to generalize that the longer the fence, the more pumpkins can be enclosed, and that as the shape of the pen gets closer to a square, more pumpkins can be enclosed. Students should also share their predictions and why they were able/unable to accurately predict the sizes of the smallest and largest patches.

8. As a class, predict the smallest and largest patch for an 18-link fence and then check the accuracy of your predictions.

Connecting Learning

1. How many pumpkins could you enclose with a 12-link fence? [5, 8, 9]

2. What were the shapes of the enclosures? [rectangles and one square]

3. Which shape enclosed the most pumpkins? [square]

4. How did these findings compare to your findings for 14- and 16-link fences?

5. Were you able to accurately predict the sizes of the smallest and largest patches for the 14-link fence? ...the 16-link fence? Why or why not?

6. What do you predict will be the number of pumpkins in the smallest patch that can be enclosed with 18 links? Why?

7. What do you predict will be the number of pumpkins in the largest patch that can be enclosed by 18 links? Why?

8. Which numbers of links made square enclosures? [12, 16] Why? [They are divisible by 4.] What do you think is the next number of links that would make a square enclosure? [20] Why?

9. Would you be able to follow our rules and enclose pumpkins using a 17-link fence? [No.] Why or why not? [You need to have an even number of paper clips to enclose pumpkins using our rules.]

Extensions

1. Eliminate the requirement that enclosures be only rectangular or square in shape. This will allow students to enclose more numbers of pumpkins for each fence length by making T-, L-, or stair-shaped enclosures.

2. Have students calculate and record the perimeter (number of paper clips) and relate that to the area (number of pumpkins enclosed). This can lead to the generalization that for a constant perimeter, the area approaches a maximum as the shape approaches a square. (This generalization applies to rectangular shapes—the largest area that can be enclosed by a given perimeter is a circle.)

3. Challenge students to identify the number of paper clips in fences that enclosed a square, and discover the next number of paper clips that will enclose a square area. These numbers can be studied for patterns that can then be generalized. (Paper clip fences that are multiples of four can enclose square areas.)

* Reprinted with permission from *Principles and Standards for School Mathematics*, 2000 by the National Council of Teachers of Mathematics. All rights reserved.

Pumpkin Patches

Key Questions

How many pumpkins can be enclosed with fences of different lengths?

Learning Goals

Students will:

1. use chains of paper clips (fences) to enclose patches of pumpkins,
2. discover the different numbers of pumpkins that can be enclosed with a fence of a given length,
3. describe the shape of each enclosure,
4. predict how many pumpkins they will be able to enclose with a larger fence, and
5. test their predictions.

12-Link Fence

Number of Pumpkins	Shape of Patch

For a 14-link fence I predict:

The smallest patch will have _____ pumpkins.

The largest patch will have _____ pumpkins.

14-Link Fence

Number of Pumpkins	Shape of Patch

For a 16-link fence I predict:

The smallest patch will have _____ pumpkins.

The largest patch will have _____ pumpkins.

16-Link Fence

Number of Pumpkins	Shape of Patch

My predictions for a 14-link fence were right/wrong because:

My predictions for a 16-link fence were right/wrong because:

Pumpkin Patches

Connecting Learning

1. How many pumpkins could you enclose with a 12-link fence?

2. What were the shapes of the enclosures?

3. Which shape enclosed the most pumpkins?

4. How did these findings compare to your findings for 14- and 16-link fences?

5. Were you able to accurately predict the sizes of the smallest and largest patches for the 14-link fence? ...the 16-link fence? Why or why not?

Pumpkin Patches

Connecting Learning

6. What do you predict will be the number of pumpkins in the smallest patch that can be enclosed with 18 links? Why?

7. What do you predict will be the number of pumpkins in the largest patch that can be enclosed by 18 links? Why?

8. Which numbers of links made square enclosures? Why? What do you think is the next number of links that would make a square enclosure? Why?

9. Would you be able to follow our rules and enclose pumpkins using a 17-link fence? Why or why not?

Topic
Problem solving

Key Question
How can we use a balance to compare the mass of two or more objects?

Learning Goals
Students will:
1. devise a plan for ordering objects from lightest to heaviest using only a balance and the objects themselves, and
2. explain the method they used using words and/or diagrams.

Guiding Document
NCTM Standards 2000*

- *Recognize the attributes of length, volume, weight, area, and time*
- *Understand how to measure using nonstandard and standard units*
- *Build new mathematical knowledge through problem solving*
- *Solve problems that arise in mathematics and in other contexts*
- *Apply and adapt a variety of appropriate strategies to solve problems*
- *Monitor and reflect on the process of mathematical problem solving*

Math
Measurement
    mass
Problem solving

Integrated Processes
Observing
Comparing and contrasting
Applying
Relating
Recording

Problem-Solving Strategies
Guess and check
Use manipulatives

Materials
Balances (see *Management 1*)
Classroom objects (see *Management 5*)
Student pages

Background Information
Measurement should be one of the key strands in any good school mathematics curriculum. It is ironic, then, that measurement is so often skipped. In the past, many math textbooks gave only lip service to measurement and included a chapter on it somewhere after the important chapters on computation. (The measurement chapter was usually not very exciting and often involved superficial activities—such as looking at pictures of rulers with objects above them and measuring how long the objects were.) More recently, the national reform documents have called for authentic measurement to be incorporated into the mathematics classroom. However, even with this push, many classrooms are not doing much with measurement. In some cases this may be due to a lack of equipment, in others a fear of the metric system, and in others the feeling that measurement just isn't as important as other parts of an already crowded curriculum. While these stumbling blocks can be understood, it is sad when nothing is done about them. The end result is that many students miss the opportunity to develop the concepts and skills of measurement. This is truly unfortunate, since measurement is an important life skill that should be a part of every school's curriculum. One of the great strengths of AIMS is that many of its activities incorporate measurement in a way that can easily be done in most classrooms. This activity is one of these.

Management
1. The activity works best if there are enough balances available for students to work in pairs or small groups.
2. If your students have not had much experience with balances, this is an excellent introductory activity. However, even those students who have had experience with balances should find the higher levels of this problem challenging and worthwhile.

3. Prior to doing the activity, demonstrate how to equalize a balance by using the slide. This is something that you will want to be sure the students apply whenever they use balances.
4. As students devise their methods of ordering the objects and work on this activity, you have an excellent opportunity to gain insight into their mathematical thinking. This insight is crucial in order to better understand how students construct their mathematical knowledge.
5. It is not necessary to gather sets of objects for students to use in this investigation; however, it is important that you have a selection of objects available in the classroom for them to use. Things such as scissors, erasers, staplers, etc., would all work well.

Procedure
1. Explain to the students that they will be doing some problem solving that will involve using a balance. Demonstrate how to equalize a balance.
2. Distribute the student pages.
3. Ask the students to find several, (at least five) objects in the classroom that would fit into the pan of the balance.
4. Direct the students' attention to *Task One* on the first student page. Prior to distributing the balances, have the students devise a method for comparing the mass of two, three, four, and five objects. When students have explained their methods using words and/or diagrams, distribute a balance to each group of students.

5. Provide time for students complete the tasks using their individual methods.
6. When the groups have completed their comparisons and have ordered their five objects from lightest to heaviest, allow them to share their methods and results with the class.

Connecting Learning
1. Which of the objects had the greatest mass? ...the least mass?
2. How did you go about determining the order of the objects without finding the actual mass of each object?
3. Why was it not important for each group to use the same objects?
4. How did you method change as you added additional objects to the original set of two?
5. Describe the task that you designed as task five.

Extensions
1. Have students find the actual mass of each object.
2. Provide a set of objects to order that are very close in mass.

* Reprinted with permission from *Principles and Standards for School Mathematics*, 2000 by the National Council of Teachers of Mathematics. All rights reserved.

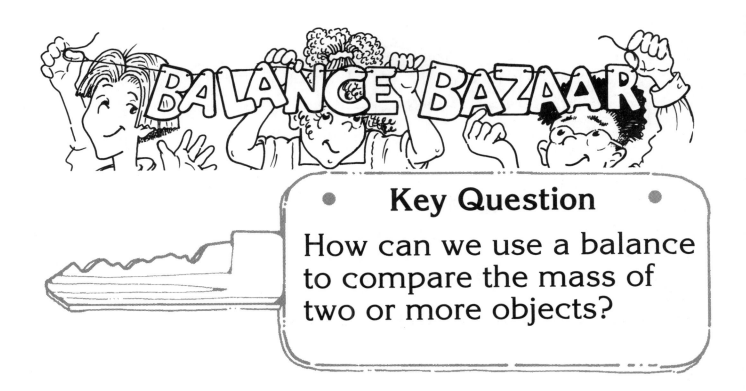

BALANCE BAZAAR

Key Question

How can we use a balance to compare the mass of two or more objects?

Learning Goals

Students will:

1. devise a plan for ordering objects from lightest to heaviest using only a balance and the objects themselves, and
2. explain the method they used using words and/or diagrams.

A balance is a measuring tool that can be used to compare the masses of two or more different objects.

Devise a plan for each the following tasks. After completing each task, explain, in words and/or diagrams, the method(s) you used.

Method:

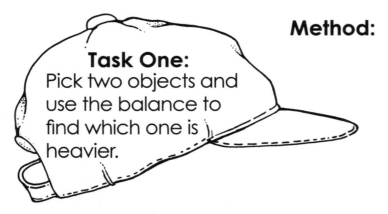

Task One:
Pick two objects and use the balance to find which one is heavier.

Method:

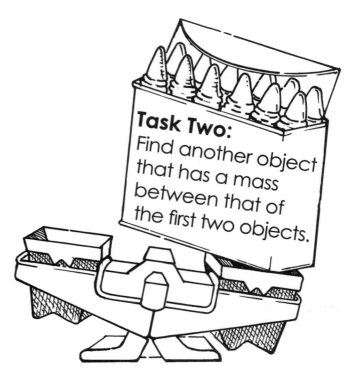

Task Two:
Find another object that has a mass between that of the first two objects.

81

BALANCE BAZAAR

Method:

Task Three:
Pick four different objects. Use the balance to arrange them from lightest to heaviest.

Method:

Task Four:
Repeat *Task Three* with five objects.

Task Five:
Devise your own task that uses the balance and describe it on the back of this paper.

Connecting Learning

1. Which of the objects had the greatest mass? …the least mass?

2. How did you go about determining the order of the objects without finding the actual mass of each object?

3. Why was it not important for each group to use the same objects?

4. How did you method change as you added additional objects to the original set of two?

5. Describe the task that you designed as task five.

SCORE KEEPERS

Topic
Problem solving

Key Question
How can you determine a set of missing numbers when given specific target sums?

Learning Goals
Students will:
1. use problem-solving skills to select numbers so that they total a target sum, and
2. gain practice using addition to solve problems.

Guiding Document
NCTM Standards 2000*
- *Develop fluency with basic number combinations for addition and subtraction*
- *Understand the effects of adding and subtracting whole numbers*
- *Develop and use strategies for whole-number computations, with a focus on addition and subtraction*
- *Build new mathematical knowledge through problem solving*
- *Apply and adapt a variety of appropriate strategies to solve problems*

Math
Number and operations
    addition
Problem solving

Integrated Processes
Observing
Comparing and contrasting
Collecting and recording data
Interpreting data

Problem-Solving Strategies
Guess and check
Use manipulatives

Materials
Light-colored crayons
Number cards (see *Management 2*)
Student pages

Background Information
In this activity, students are asked to find one solution out of many possible answers. They will be asked to manipulate a set of number cards to solve a series of word problems. Using the guess and check strategy, they can try out different solutions until they find the one that meets the criteria. As students guess and eliminate possible solutions, they will get closer to the correct answer.

Management
1. This set of problems is designed to allow students to practice addition at the problem-solving level. Use it multiple times throughout the year for extended practice.
2. Students will need number cards labeled with the digits zero through nine. These cards can be small sticky notes, scratch paper cut into small squares, or numbered math chips.
3. There are several different problems, which vary in difficulty based on the number of missing digits. Choose the problems that are most appropriate for your students. It is suggested that you use the marble problem to introduce the students to the guess and check strategy, since it is the easiest problem in the series.

Procedure
1. Give each student one of the student pages, a crayon, and a set of number cards.
2. Direct the students to highlight the target sums in crayon. For example, the possible target sums for the marble puzzle are 12, 15, and 18.
3. Challenge the students to choose the correct numbers from the set of number cards that can total all of the target sums. Discuss the students' solutions and the strategies they used.
4. Continue assigning the remaining sports problems. Close each lesson with a time of class discussion in which students look for patterns in their solutions and describe the strategies they used to solve the problems.

Connecting Learning
1. What are the two rings worth in the marble game? [6 points, 9 points] How do you know?
2. How many points is each circle worth on the dartboard? [3 points, 5 points, 7 points] How do you know?
3. What are the numbers in the skee ball game? [2, 4, 6, 8] How do you know?
4. What did you do that helped you solve the problems?
5. How does your strategy compare to the strategies used by your classmates?
6. Were some problems easier to solve? ...harder to solve? Why?
7. What number patterns did you notice?
8. Were the target sums always even numbers, always odd numbers, or a combination of both odd and even numbers?

SCORE KEEPERS

Learning Goals

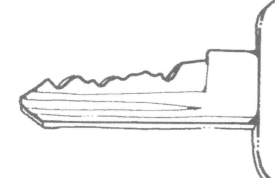

Students will:

1. use problem-solving skills to select numbers so that they total a target sum, and
2. gain practice using addition to solve problems.

SCORE KEEPERS

MARBLE PROBLEM

Juan has just joined a marble game in progress. There are two rings drawn on the ground, and each player has shot three marbles. He hears the totals of the players are 12, 15, and 18. How many points is each ring worth?

SCORE KEEPERS

DART PROBLEM

You get 3 darts, and the following totals are possible 9, 11, 13, 15, 17, 19, 21.

How many points is A worth?

How many points is B worth?

How many points is C worth?

SCORE KEEPERS

SKEE BALL PROBLEM

Charles just got a skee ball game for his birthday. The game has four scoring areas and four balls, but the numbers on his game are missing. The possible totals that he can make are 8, 10, 12, 14, 16, 18, 20, 22, 24, 26, 28, 30, 32, and 36. What numbers are on Charles' game?

Connecting Learning

1. What did you do that helped you solve the problems?

2. How does your strategy compare to the strategies used by your classmates?

3. Were some problems easier to solve? …harder to solve? Why?

4. What number patterns did you notice?

5. Were the target sums always even numbers, always odd numbers, or a combination of both odd and even numbers?

Problem-Solving Strategies
Organize the Information

It is often helpful to organize the information when trying to solve problems. You can put what you know into a list, chart, or table. Then you can see what you still need to solve the problem. You can also use this strategy when you need to find lots of different solutions.

PROBABLY BEARS

Topic
Probability

Key Questions
1. How can organizing information help us?
2. How many bears of each color are in the cave?

Learning Goals
Students will:
1. design an organization scheme so that they can keep track of the type and number of bears pulled in four random samples, and
2. predict what the population of bears in the cave looks like based on the samples.

Guiding Document
NCTM Standards 2000*
- *Predict the probability of outcomes of simple experiments and test the predictions*
- *Describe events as likely or unlikely and discuss the degree of likelihood using such words as certain, equally likely, and impossible*
- *Collect data using observations, surveys, and experiments*
- *Represent data using tables and graphs such as line plots, bar graphs, and line graphs*
- *Build new mathematical knowledge through problem solving*

Math
Probability
Problem solving

Integrated Processes
Observing
Classifying
Predicting
Collecting and recording data
Comparing and contrasting
Interpreting data
Drawing conclusions
Generalizing

Problem-Solving Strategies
Organize the information
Use manipulatives

Materials
24 Teddy Bear Counters (see *Management 1*)
Small brown lunch bag

Background Information
It is often necessary for students to organize information in order to solve problems. In many cases, when we expect students to organize data, we as teachers provide them with the organization scheme and they simply fill in the information as it is gathered. This activity is designed to be an open-ended opportunity for students to determine the best way to organize the information they are expected to gather. It will also give the students an opportunity to explore probability concepts. The students will be told that there are four different colors of bears, 24 total bears, hidden in a cave. They will be allowed to take four random samples (a sample in which no bias occurs in the selections) and will be asked to predict what the population (the entire group) looks like based on the samples.

Management
1. Prior to doing this activity, place one yellow bear, 12 green bears, six red bears and five blue bears in a brown paper lunch bag.
2. After each sample is counted and recorded, place the bears back into the bag.

Procedure
1. Tell the students that you have a brown cave (lunch bag) with 24 bears in it. Explain that there are four different colored bears inside and that it will be their job to predict how many bears of each color are in the cave.
2. Tell the students that you would like the predictions to be based on some information and not just guesses. Encourage the students to suggest ways that they could get information about the bears in the bag without looking inside. If it is not mentioned, suggest that you pull out a few random samples.
3. Tell the students that you will be pulling out four random samples of six bears on which they can base their predictions.

4. Ask them how organizing the data will help them with their predictions. Discuss what kinds of information they will be getting from the random samples, such as color and number of bears for sample one, two, three and four, etc.

5. Provide a few minutes for the students to decide how they want to set up their organization schemes.

6. Pull six bears out of the bag and tell the children how many of each color are in the sample. Place the bears on display so that the students can refer to the bear counters as they record the information. Repeat this process three more times. Allow time for the students to record the information after each sample is pulled.

7. Ask the students to study their data and predict how many bears of each color are in the cave. Remind the students that the total number of bears in the cave is 24. Therefore, when they add the predicted number of blue, red, yellow, and green bears, it should equal 24.

8. Pour the bears out of the cave and record the actual number of each color bear. Have the students compare the actual count with their predictions. Discuss the different ways that the students organized the information they received from the samples. Discuss how a good organization scheme helped with the predictions.

Connecting Learning

1. Why is it important to place the bears back into the bag after each random sample is taken?

2. How did you keep track of what was pulled out of the bag each time? Explain.

3. Why is it important to organize information in an activity like this?

4. How did your prediction compare to the actual contents of the cave?

5. Where else might random sampling be used?

Extension

Have the students set up their own cave full of hidden bears and design a student worksheet complete with organization chart for another class to use.

* Reprinted with permission from *Principles and Standards for School Mathematics*, 2000 by the National Council of Teachers of Mathematics. All rights reserved.

PROBABLY BEARS

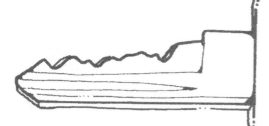

Key Questions

1. How can organizing information help us?
2. How many bears of each color are in the cave?

Learning Goals

Students will:

1. design an organization scheme so that they can keep track of the type and number of bears pulled in four random samples, and
2. predict what the population of bears in the cave looks like based on the samples.

Welcome
Please wipe your paws!

PROBABLY BEARS

How many bears of each color are in the cave?

There are 24 bears hidden in the cave. There are four different colors of bears.

Organize the Information

	PREDICTION	ACTUAL
Yellow		
Green		
Blue		
Red		
TOTAL	24	24

Connecting Learning

1. Why is it important to place the bears back into the bag after each random sample is taken?

2. How did you keep track of what was pulled out of the bag each time? Explain.

3. Why is it important to organize information in an activity like this?

4. How did your prediction compare to the actual contents of the cave?

5. Where else might random sampling be used?

Topics
Geomerty: coordinate grid

Key Questions
1. How can we use a coordinate system to locate counting bears?
2. How can we keep track of the number of guesses it takes to hunt down the bear?

Learning Goals
Students will:
1. locate the positions of bears on a coordinate grid,
2. use problem-solving strategies to locate the bears in as few guesses as possible, and
3. design an organizational scheme that will allow them to keep track of the number of guesses it took them to hunt down the bear.

Guiding Document
NCTM Standards 2000*
- *Make and use coordinate systems to specify locations and to describe paths*
- *Represent data using tables and graphs such as line plots, bar graphs, and line graphs*
- *Collect data using observations, surveys, and experiments*
- *Build new mathematical knowledge through problem solving*
- *Solve problems that arise in mathematics and in other contexts*
- *Apply and adapt a variety of appropriate strategies to solve problems*
- *Monitor and reflect on the process of mathematical problem solving*

Math
Geometry
    coordinate grid
Problem solving

Integrated Processes
Observing
Collecting and recording data
Interpreting data
Drawing conclusions
Generalizing

Problem-Solving Strategies
Organize the information
Use logical thinking

Materials
For each pair:
    two Teddy Bear Counters
    two copies of the coordinate grid
    white copy paper
    divider (see *Management 3*)

For the class:
    transparency of grid

Background Information
Being able to organize information is an important problem-solving strategy. One way of organizing information is by making a list of the information. This helps students see what is known as well as what information they still need to find out. Tables and graphs are other useful ways to organize the information you have. Tables and graphs help you keep track of data.

In this lesson, students will find an organized way to keep track of the number of guesses it takes to hunt down each bear as well as apply the use of a coordinate system to locate the bears. The lesson is an application for the coordinate grid. Therefore, it is assumed that the children already know how to locate points on a coordinate grid and name them by calling out the ordered pairs, such as (2, 4).

Management
1. Students should already know how to read the axes on a coordinate grid. Remind students to read the horizontal or *x* axis first, followed by the vertical or *y* axis when identifying coordinate points. Remind them to first go over and then go up.
2. When placing the bears on the grid paper, place them on the intersections of lines, not in the spaces between lines.
3. Each pair of students will need something such as a textbook or folder to create a wall between them. It is important that they not be able to see their partner's grid paper.

Procedure

1. To review how to use a coordinate grid, lay the transparency of the grid onto the overhead. Place a Teddy Bear Counter at a specific point where two lines intersect on the grid, and ask the students to name the location. For example they may tell you that the bear is at (2, 3).

2. Explain to the students that they will be going bear hunting on a coordinate grid. Tell them that they will be playing this game in partners and that the object of the game is to hunt down the bear in the fewest number of guesses.

3. Give each student a Teddy Bear Counter, a sheet of grid paper, and a blank piece of white paper.

4. Have the students sit across from each other with a divider between them so that they cannot see the other person's grid.

5. Ask both partners to secretly hide their bears at an intersection on the grid.

6. Remind the students that since the object of the game is to hunt down the bear in the fewest number of guesses, they will need to design a way to keep track of the number of guesses each player takes. Allow time for the partners to come up with a plan for organizing the information.

7. Have the students play one round of bear hunting by taking turns guessing coordinates. Play continues until both partners have located the other player's bear.

8. Talk about the different methods the students used to keep track of the number of guesses each player used before hunting down the bear. Ask the students to evaluate the effectiveness of their organizational plans.

Connecting Learning

1. How did you keep track of the number of guesses it took you to hunt down the bears? Is there a better way?

2. How did you begin your search for the bear?

3. Describe one position where you located a bear.

4. How did the coordinate grid help you find the location of the bears?

5. How will you change your hunting strategy when you play this game again?

* Reprinted with permission from *Principles and Standards for School Mathematics*, 2000 by the National Council of Teachers of Mathematics. All rights reserved.

Key Questions

1. How can we use a coordinate system to locate counting bears?
2. How can we keep track of the number of guesses it takes to hunt down the bear?

Learning Goals

Students will:

1. locate the positions of bears on a coordinate grid,
2. use problem-solving strategies to locate the bears in as few guesses as possible, and
3. design an organizational scheme that will allow them to keep track of the number of guesses it took them to hunt down the bear.

Bear Hunt

Hide your bear at any intersection of the grid. Keep track of the guesses your partner makes. Count how many guesses it takes to find the bear.

	1	2	3	4	5	6	7
5							
4							
3							
2							
1							
0							

Bear Hunt

Connecting Learning

1. How did you keep track of the number of guesses it took you to hunt down the bears? Is there a better way?

2. How did you begin your search for the bear?

3. Describe one position where you located a bear.

4. How did the coordinate grid help you find the location of the bears?

5. How will you change your hunting strategy when you play this game again?

CA$H COMBOS

Topic
Combinations

Key Question
How many ways can you make $50 using any combination of $5, $10, and/or $20 bills?

Learning Goals
Students will:
1. find as many ways as they can to make $50 using combinations of smaller bills, and
2. develop a way to organize their solutions to determine if they have found all of the possibilities.

Guiding Document
NCTM Standards 2000*
- *Represent data using tables and graphs such as line plots, bar graphs, and line graphs*
- *Build new mathematical knowledge through problem solving*
- *Create and use representations to organize, record, and communicate mathematical ideas*

Math
Whole number operations
    addition
Problem solving

Integrated Processes
Observing
Organizing
Recording
Comparing and contrasting

Problem-Solving Strategy
Organize the information

Materials
Student page

Background Information
This activity asks students to find as many ways as they can to make $50 using any combination of $5, $10, and/or $20 bills. (There are 12 ways.) While this is not a difficult task, the multiple combinations may prove unwieldy for some students unless they devise a systematic method of recording them. Without some system, it is hard to tell if all possible combinations have been found.

Management
1. Two different versions of *Cash Combos* are provided; you will need to determine which one is most appropriate for your students. The first version is open-ended and designed to allow students to develop their own systems of recording information. The second version is also open-ended, but adds a bit more structure by providing a blank table in which students can record their combinations. They will still have to devise a system to determine if they have found all the combinations.

Procedure
1. Have students get into small groups and distribute the desired student page to each student.
2. Give the groups time to work on the problem and come up with a variety of solutions.
3. Once groups have developed a way to organize their solutions and are confident they have found all of the possibilities, close with a time of discussion where groups share their answers and explain the methods they used to ensure that all of the solutions were discovered.

Connecting Learning
1. How many ways are there to make $50 using combinations of $5, $10, and/or $20 bills? [12]
2. How do you know that you have found all of the solutions?
3. How does the method your group developed compare to the methods of other groups?
4. Do you think one method is easier than another? Why or why not?

Extension
Add $1 bills to the problem. This minor addition increases the number of combinations from 12 to over 50.

* Reprinted with permission from *Principles and Standards for School Mathematics*, 2000 by the National Council of Teachers of Mathematics. All rights reserved.

CA$H COMBOS

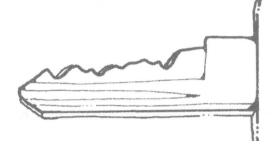

Key Question

How many ways can you make $50 using any combination of $5, $10, and/or $20 bills?

Learning Goals

Students will:

1. find as many ways as they can to make $50 using combinations of smaller bills, and
2. develop a way to organize their solutions to determine if they have found all of the possibilities.

CA$H COMBOS

In how many different ways can you make a total of $50 using any combination of $5, $10, and/or $20 bills?

List your combinations below.

Do you have all the possible combinations?
How do you know?

If necessary, use the back of this paper to redo your list in such a way that you are *sure* you have found all the combinations.

CA$H COMBOS

In how many different ways can you make a total of $50 using any combination of $5, $10, and/or $20 bills?

List your combinations in the table below.

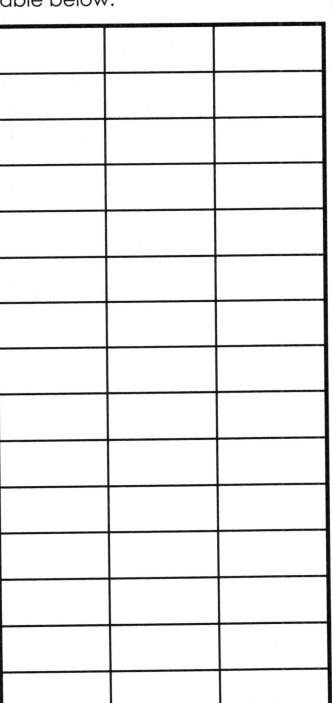

Do you have all the possible combinations?
How do you know?

If necessary, use the back of this paper to redo your list in such a way that you are *sure* you have found all the combinations.

CA$H COMBOS

Connecting Learning

1. How many ways are there to make $50 using combinations of $5, $10, and/or $20 bills?

2. How do you know that you have found all of the solutions?

3. How does the method your group developed compare to the methods of other groups?

4. Do you think one method is easier than another? Why or why not?

Problem-Solving Strategies
Look for Patterns

Patterns are everywhere. Some patterns repeat. Other patterns grow. Some patterns use numbers. Some patterns use shapes. Some patterns are cycles. Looking for patterns can help you solve problems. Sometimes, knowing the pattern and how to extend it gives you the answer.

Pattern Play

Topic
Patterns

Key Question
How can identifying patterns help us solve problems?

Learning Goal
Students will describe, extend, and make generalizations about geometric and numeric patterns.

Guiding Documents
Project 2061 Benchmark
- *Patterns can be made by putting different shapes together or taking them apart.*

NCTM Standards 2000*
- *Describe, extend, and make generalizations about geometric and numeric patterns*
- *Represent and analyze patterns and functions, using words, tables, and graphs*
- *Build new mathematical knowledge through problem solving*
- *Solve problems that arise in mathematics and in other contexts*
- *Apply and adapt a variety of appropriate strategies to solve problems*

Math
Patterns
Problem solving

Integrated Processes
Observing
Comparing and contrasting
Classifying
Predicting

Problem-Solving Strategies
Look for patterns
Use manipulatives
Use logical thinking

Materials
Transparencies (see *Management 4*)
Pattern cards
1-inch grid paper
Color counters (see *Management 7*)
1 roll of pennies

Background Information
   Looking for patterns helps bring order and predictability to what may otherwise seem like chaos. The ability to recognize patterns and functional relationships is a powerful problem-solving tool. It allows students to simplify tasks and to make generalizations beyond the information given.

   In this activity, students will analyze and make generalizations about a variety of patterns using t-tables, spatial sense, and logical thinking to solve problems. The patterns included in this activity will address number concepts, algebra, geometry, measurement, and data organization skills. The intent is to introduce students to the underlying ideas of mathematics through problem-solving scenarios. The problems will require the students to apply simple knowledge of arithmetic as they devise plans to solve the problems.

Management
1. It is expected that the children have had prior pattern experiences that include the use of t-tables or function machines.
2. The pattern task cards included in this lesson can be used as station cards, book pages for individual student books, or one at a time as a beginning to math class. Copy sets of the pattern task cards onto card stock and laminate for extended use.
3. As students gain an understanding of color and shape patterns, ask them questions that focus on numbers. For example, "What would appear in the sixth position of the pattern? ...the ninth position? ...the 12th position?"
4. Make overhead transparencies of each of the pattern cards.
5. To solve each problem, students may need various manipulatives. They can use manipulatives from the classroom such as rulers, Unifix cubes, etc.
6. Make several copies of the grid paper and make them available for students solving pattern two.

7. Each group or student will need a collection of red, green, and blue counters for the fifth pattern—at least 10 of each color. These counters could be Unifix cubes, colorful goldfish crackers, bingo chips, etc.

Procedure
1. Review the different types of patterns. [e.g., growing, repeating, number] For example, use colored blocks to create patterns and have students determine what comes next. Ask them to share how they know their solutions are correct. Provide number patterns such as 88, 77, 66, and ask what comes next. Also, use a hundred's chart and look for a variety of patterns (counting by fives, twos etc.; using diagonals, horizontal and vertical lines on the hundred's chart to find the patterns).
2. Use the transparencies suggested in *Management 4* to introduce each individual pattern card.
3. After introducing the pattern cards, use them at math centers or begin each class with a different pattern problem.
4. End each problem with a discussion of the solution and strategies used by the students.

Connecting Learning
1. How did you solve each problem?
2. Which clues were the most helpful in *Noodlin' With Numbers*? Why?
3. What, if anything confused you?
4. What math skills did you have to use to solve these problems?
5. What else could be used instead of numbers in the *Sudoku* puzzle?
6. When in the real world do you have to look for patterns to solve problems?

Solutions
Noddlin' With Numbers—Pattern One
The number villain is doubling each number and adding one. ($2x + 1$)

Net Sense—Pattern Two
A is the correct net. The missing numbers are 1, 2, and 2.

Sudoku—Pattern Three

1	4	3	6	2	7	9	8	5
8	2	5	1	9	4	3	6	7
9	7	6	3	5	8	2	4	1
5	3	9	7	4	1	6	2	8
2	6	7	9	8	3	1	5	4
4	8	1	5	6	2	7	3	9
6	1	4	2	7	5	8	9	3
3	9	8	4	1	6	5	7	2
7	5	2	8	3	9	4	1	6

Who is missing?—Pattern Four
The missing box should be four centimeters wide and six centimeters tall.

School's In—Pattern Five
The last fish in line is red.
There are eight blue fish in the school

Pyramid of Pennies—Pattern Six
There are 19 pennies in the 10th row.
The 14th row had 27 pennies.

*   Reprinted with permission from *Principles and Standards for School Mathematics*, 2000 by the National Council of Teachers of Mathematics. All rights reserved.

Pattern Play

Key Question

How can identifying patterns help us solve problems?

Learning Goal

Students will:

describe, extend, and make generalizations about geometric and numeric patterns.

Noodlin' With Numbers—Pattern One

Pattern detectives everywhere, we need your help! A dreadful number villain is vandalizing our town. No number is safe from his evil clutches.

This mysterious phantom is noodlin' with our numbers and causing chaos in our town. You can imagine the confusion when highway 3 suddenly became highway 7 and the speed limit changed from 20 to 41 miles an hour within minutes. Addresses are being changed. The ice cream shop isn't 127 Cedar Avenue anymore. It is now 255 Cedar Avenue.

We're stumped. Please help us solve this mystery. Do you see a pattern? Maybe if we can discover his pattern, we can save the next town.

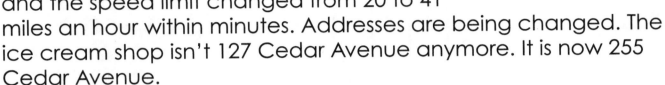

Net Sense—Pattern Two

Which net will fold to make the cube pictured? What numbers would appear where the question marks are?

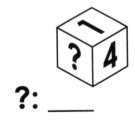

?: _____   ?: _____   ?: _____

A

	6	
	5	
2	1	3
	4	

B

	3
6	2
	5
1	4

C

		1
6		2
5	3	4

Sudoku—Pattern Three

Fill in the grid so that every row, every column, and every 3 x 3 box contains the digits 1 through 9.

1	4	3						5
8			1	9	4			
			3			2	4	1
5		9				6	2	
	6	7		8	3			
			5			7		9
			2	7			9	3
3	9			1	6			
	5	2				4	1	

Who is Missing?—Pattern Four

Measure the boxes and draw the one that should go in between.

School's In—Pattern Five

There are 20 fish in the "school." There are three different colors of fish—blue, green, and red. The school was lined up so that the pattern of fish was blue, blue, green, green, red, blue, blue, green, green, red...

- What color is the last fish in the line?

- How many blue fish are in the school?

Pyramid of Pennies—Pattern Six

Chin built a pyramid of pennies. He had one penny on top. He had three pennies in the second row. He had five pennies in the third row and seven pennies in the fourth row.

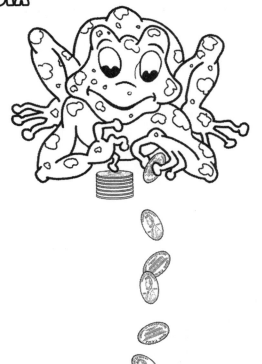

- How many pennies did he have in the 10th row?

- Which row had 27 pennies?

113

Pattern Play

Connecting Learning

1. How did you solve each problem?

2. Which clues were the most helpful in *Noodlin' With Numbers?* Why?

3. What, if anything confused you?

4. What math skills did you have to use to solve these problems?

5. What else could be used instead of numbers in the *Sudoku* puzzle?

6. When in the real world do you have to look for patterns to solve problems?

Snappy Patterns

Topic
Problem solving

Key Question
How many different patterns can you make with two or more colors of Unifix cubes?

Learning Goals
Students will:
1. use fractions to determine the number and colors of Unifix cubes they will be using,
2. make as many patterns as they can with those Unifix cubes,
3. record these patterns, and
4. translate the patterns into letters.

Guiding Document
NCTM Standards 2000*
- *Recognize, describe, and extend patterns such as sequences of sounds and shapes or simple numeric patterns and translate from one representation to another*
- *Describe, extend, and make generalizations about geometric and numeric patterns*
- *Understand and represent commonly used fractions such as ¼, ⅓, and ½*
- *Apply and adapt a variety of appropriate strategies to solve problems*
- *Solve problems that arise in mathematics and in other contexts*
- *Build new mathematical knowledge through problem solving*

Math
Number and operations
  fractions
Patterns
Problem solving

Integrated Processes
Observing
Organizing
Collecting and recording data
Analyzing

Problem-Solving Strategies
Look for patterns
Use manipulatives

Materials
Unifix cubes (see *Management 1*)
Colored pencils or crayons (see *Management 2*)
Transparency of recording page
Student page

Background Information
Patterns are everywhere in the world around us, and learning to create and recognize them is a valuable skill for students to have. This activity allows students to explore repeating patterns with Unifix cubes. These patterns are then translated into letters, which allows them to be analyzed for repetition.

Management
1. Each group of students will need Unifix cubes in six different colors. They will need six each of two colors, and should have at least four each of the other colors.
2. The colored pencils or crayons should match the colors of Unifix cubes available to students so that they can record their patterns.
3. You will need a transparency of the class recording page and colored overhead pens with which to record the class solutions.
4. It is expected that students have had practice generating repeating patterns prior to this experience.

Procedure
1. Have students get into groups and distribute the Unifix cubes, colored pencils, and several copies of the student page.
2. If necessary, review some different kinds of repeating patterns and how they are made.
3. Ask the groups to select 12 Unifix cubes so that half are one color and half are another color. Circulate among the groups to verify that everyone has chosen the correct number of cubes of each color.
4. Tell students that their challenge is to use all 12 Unifix cubes to make as many different repeating pattern trains as they can in five minutes. Each of these patterns must be recorded on the student page. (Have additional student pages available in case groups run out of room.)

5. Inform students that you must be able to see at least one entire repetition of the pattern for it to be valid. (This means that all valid patterns will repeat in twos, threes, fours, or sixes.)

6. After groups have had time to find several solutions, tell them to stop recording. Ask each group to share one pattern that they discovered until all of the different patterns discovered have been shared. (At this point, put up patterns that are the same except for the order of the colors.)

7. Record each of the patterns shared by the groups on the overhead transparency of the class recording page. To prevent confusion, use the same two colors for each group's patterns, even if they actually used different colors. As a class, translate the color patterns into letters (ABAB, AABBAABB, etc.).

8. Ask the students if any of the patterns that they thought were different are actually the same. If there are any duplicates, eliminate those from the recorded list.

9. As a class, see if you can find any more unique patterns using the two colors of Unifix cubes.

10. Repeat this process with more colors and other fractional color combinations such as: $\frac{1}{3}$, $\frac{1}{3}$, $\frac{1}{3}$; $\frac{2}{3}$, $\frac{1}{3}$; $\frac{1}{2}$, $\frac{1}{3}$, $\frac{1}{6}$; and $\frac{1}{3}$, $\frac{1}{3}$, $\frac{1}{6}$, $\frac{1}{6}$.

Connecting Learning

1. How many Unifix cubes of each color did you need to have half and half? [six of each]
2. How many patterns did your group discover?
3. Were any of these patterns duplicates? How do you know? [When the patterns are translated to letters, duplicates are easily spotted.]
4. Were you able to discover any more patterns as a class? Do you think you have found them all? Why or why not?
5. Which combination of Unifix cubes gave you the most different patterns? Why do you think this is?
6. What are you wondering now?

Extensions

1. Change the total number of Unifix cubes to change the lengths of repeating patterns that will work.
2. Use a single color of Unifix cubes and form stacks to make growing patterns.

*   Reprinted with permission from *Principles and Standards for School Mathematics*, 2000 by the National Council of Teachers of Mathematics. All rights reserved.

Snappy Patterns

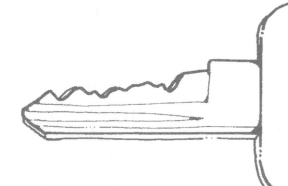

Key Question

How many different patterns can you make with two or more colors of Unifix cubes?

Learning Goals

Students will:

1. use fractions to determine the number and colors of Unifix cubes they will be using,
2. make as many patterns as they can with those Unifix cubes,
3. record these patterns, and
4. translate the patterns into letters.

Snappy Patterns

Record each pattern you find.

Class Recording Page

Snappy Patterns

Connecting Learning

1. How many Unfix cubes of each color did you need to have half and half?

2. How many patterns did your group discover?

3. Were any of these patterns duplicates? How do you know?

4. Were you able to discover any more patterns as a class? Do you think you have found them all? Why or why not?

5. Which combination of Unifix cubes gave you the most different patterns? Why do you think this is?

6. What are you wondering now?

WHAT'S MY RULE?

Topic
Patterns

Key Question
How can you make patterns with, sort, and/or order the geometric shapes you have?

Learning Goal
Students will perform a variety of tasks—as directed by the teacher—using a set of geometric shapes, including patterning, sorting, and ordering.

Guiding Documents
Project 2061 Benchmark
* *Patterns can be made by putting different shapes together or taking them apart.*

NCTM Standards 2000*
* *Sort, classify, and order objects by size, number, and other properties*
* *Describe, extend, and make generalizations about geometric and numeric patterns*
* *Analyze how both repeating and growing patterns are generated*
* *Recognize, name, build, draw, compare, and sort two- and three-dimensional shapes*
* *Build new mathematical knowledge through problem solving*

Math
Patterns
Sorting
Ordering
Geometry
   properties of shapes
Problem solving

Integrated Processes
Observing
Comparing and contrasting
Identifying

Problem-Solving Strategies
Look for patterns
Use manipulatives
Organize the information

Materials
Colored card stock—red, yellow, blue, green
Scissors

Background Information
   Open-ended problems provide great opportunities for exploring a variety of topics that are relevant to your students. They also allow for differentiation. This activity provides a wide range of suggestions for using a simple manipulative—colored paper shapes—to explore topics from patterns, to sorting, to ordering. The difficulty level and topics covered can be adjusted to fit individual students, and the same shapes can be revisited throughout the year in a variety of different ways. These experiences allow students to use a variety of problem-solving strategies while developing important content knowledge about shapes and their properties, patterns, and sorting.

Management
1. This activity is very open-ended and must be entirely teacher-directed. The manipulatives are provided, and examples of things to do are given, but the direction and structure the activity takes will be up to you as the teacher.
2. There are 16 shapes provided as manipulatives—four each of triangles, quadrilaterals, pentagons, and hexagons. Both regular and irregular shapes have been included. Each shape has been labeled with a color—red, yellow, green, or blue. There is one of each shape in each color. These shapes should be copied onto the appropriate color of card stock and cut out for students. If desired, you may laminate for durability.
3. You may wish to make a set of shapes for yourself on an overhead transparency so that you can give examples for the whole class.
4. To make the activity simpler, use fewer than the 16 shapes provided.
5. The shape sets can be used in a variety of ways. They can be placed at a center with task cards for students to explore, they can be distributed to groups and used in a whole-class setting, or they can be given to each student and used for different problems throughout the year. Use the suggestions provided in the *Procedure* to explore the topics relevant to your students.

Procedure
Part One—Patterning
1. Have students create or extend pre-created patterns based on color. The simplest of these might be a two-color pattern such as red, yellow, red, yellow…. More complex color patterns could use three or four colors (e.g., blue, green, yellow, red, blue… or yellow, green, yellow, blue, yellow, red, yellow…).

2. Challenge students to create patterns based on size, number of sides/corners, or any combination of these features. For each pattern they create or extend, students should always be able to clearly state the rule used.

Ideas for using patterns in your class:

- Glue the shapes to a sentence strip or clip them from a string hung across the chalkboard. Students can use their shapes to duplicate and extend the pattern at their desks.
- Create task cards with a variety of sample patterns drawn on them. Provide spaces in which students can extend the pattern and record the rule. Place these cards at a center.
- Have students create patterns at their desks and trade them with classmates to see if the patterns can be extended and the rules discovered.
- Have students record patterns they create along with a statement defining the rule used for that pattern.

Part Two—Sorting

1. Have students sort the shapes using a variety of rules. Shapes can be sorted by color, by number of sides/corners, by size, by angle, etc. Shapes can be sorted into two or more categories. For each sort that students perform, they should always be able to clearly state the rule used.

Ideas for presenting sorting to your class:

- Perform various sorts on the overhead. Challenge students to determine the rule and place the remainder of the pieces where they belong.
- Create a sorting mat for students and have them use it to do their own sorts.
- Challenge students to find a rule to sort the shapes into two groups, three groups, four groups, and five groups.
- Have students sort into a Venn diagram where some categories overlap.
- Have students record the sorts they make along with a statement defining the rule used for that sort.

Part Three—Ordering

1. Have students order the shapes in as many ways as they can. Shapes can be ordered from fewest to most sides, smallest to largest, and so on. For each ordering that students create, they should always be able to clearly state the rule used.

Ideas for presenting ordering to you class:

- Glue the shapes to a sentence strip or clip them from a string hung across the chalkboard. Students can use their shapes to duplicate and extend the order at their desks.
- Create task cards with a variety of sample orderings drawn on them. Provide spaces in which students can extend the order and record the rule. Place these cards at a center.
- Have students create orderings at their desks and trade them with classmates to see if the orders can be extended and the rules discovered.
- Have students record orders along with a statement defining the rule used.

Connecting Learning

1. What kinds of patterns were you able to create using the shapes?
2. What kinds of patterns are easy to create and recognize? Why?
3. What kinds of patterns are more difficult to create and recognize? Why?
4. What were some of the different ways you sorted your shapes?
5. How did your sorts compare to your classmates'?
6. What were some of the different ways you ordered your shapes?
7. How did your orders compare to your classmates'?

* Reprinted with permission from *Principles and Standards for School Mathematics,* 2000 by the National Council of Teachers of Mathematics. All rights reserved.

WHAT'S MY RULE?

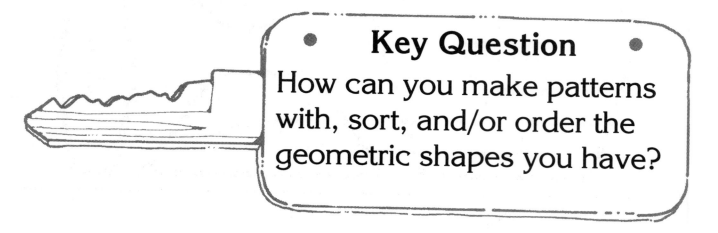

Key Question

How can you make patterns with, sort, and/or order the geometric shapes you have?

Learning Goal

Students will:

perform a variety of tasks—as directed by the teacher—using a set of geometric shapes, including patterning, sorting, and ordering.

WHAT'S MY RULE?

Copy this page onto yellow card stock. Each student needs one set of yellow shapes.

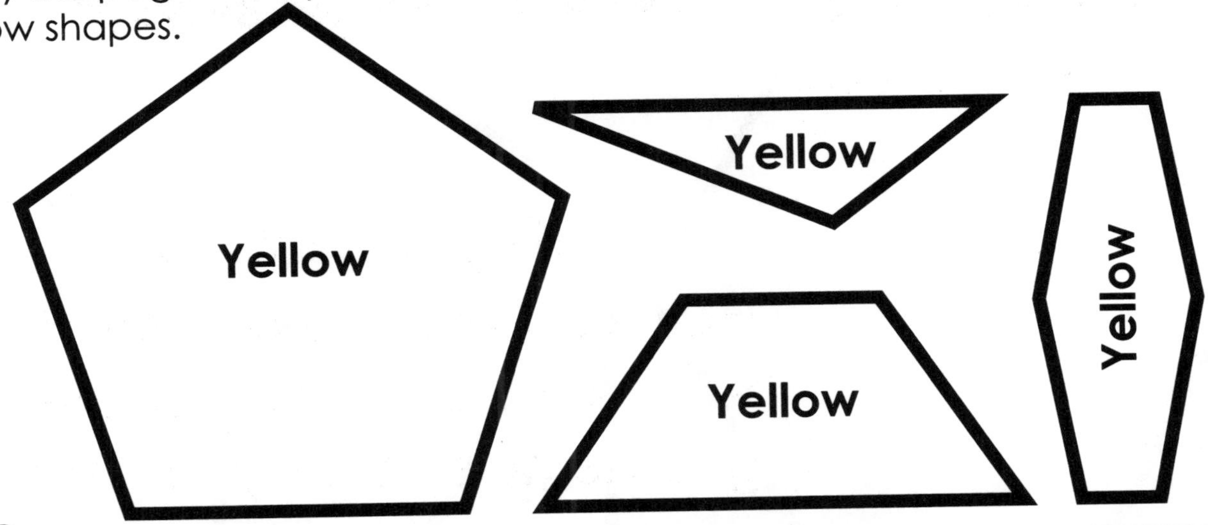

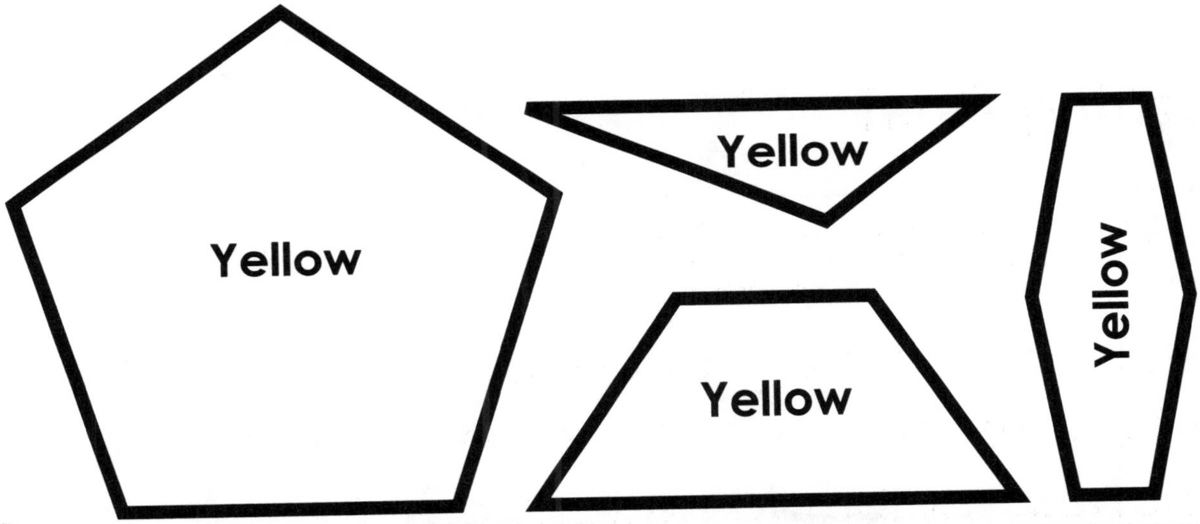

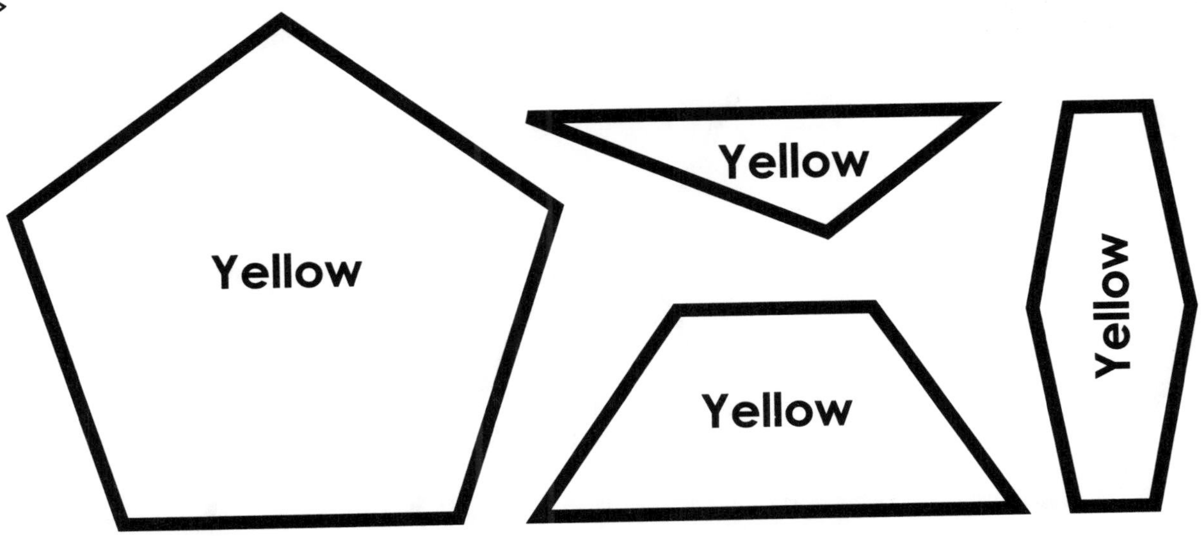

Yellow

Yellow

Yellow

Yellow

Yellow

Yellow

Yellow

Yellow

Yellow

Yellow

Yellow

Yellow

Yellow

Yellow

Yellow

WHAT'S MY RULE?

Copy this page onto red card stock. Each student needs one set of red shapes.

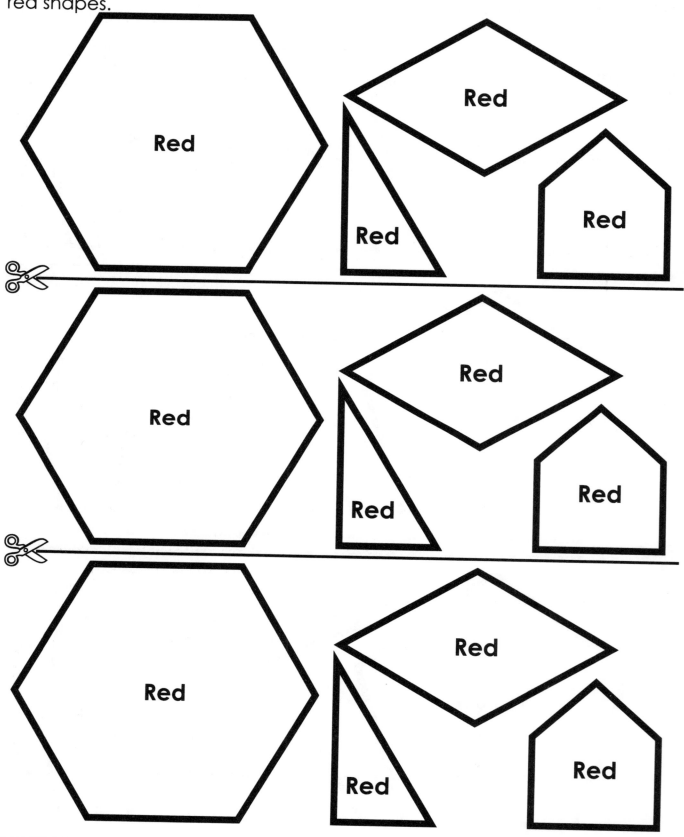

WHAT'S MY RULE?

Copy this page onto blue card stock. Each student needs one set of blue shapes.

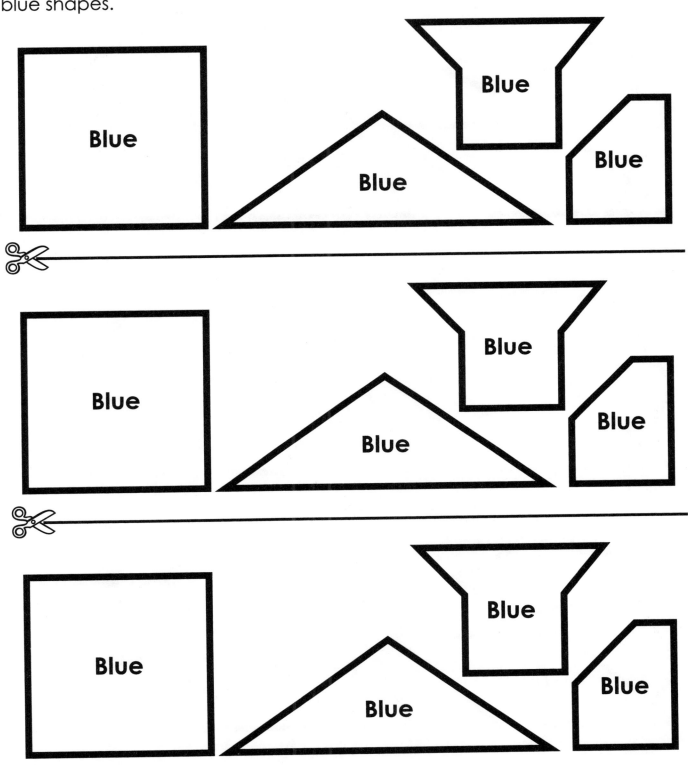

WHAT'S MY RULE?

Copy this page onto green card stock. Each student needs one set of green shapes.

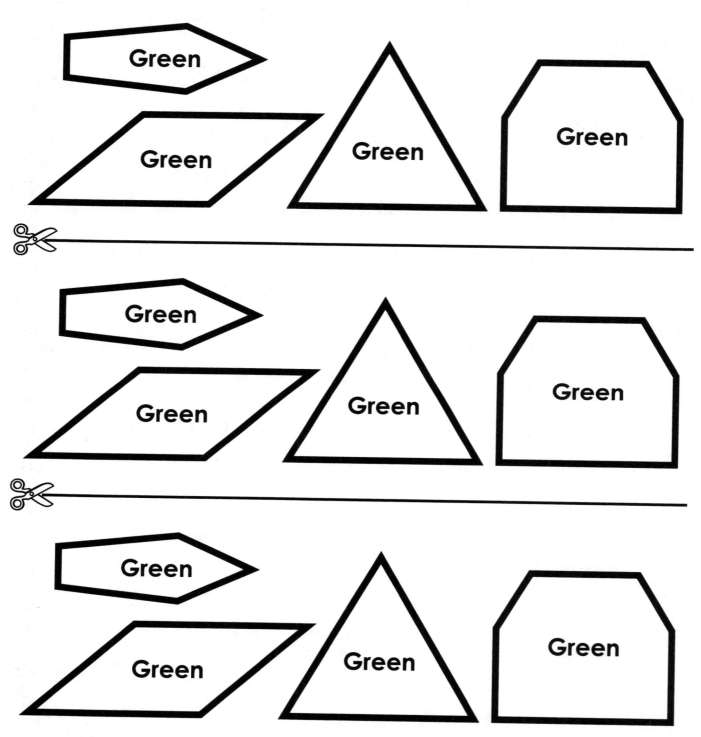

WHAT'S MY RULE?

Connecting Learning

1. What kinds of patterns were you able to create using the shapes?

2. What kinds of patterns are easy to create and recognize? Why?

3. What kinds of patterns are more difficult to create and recognize? Why?

4. What were some of the different ways you sorted your shapes?

5. How did your sorts compare to your classmates'?

6. What were some of the different ways you ordered your shapes?

7. How did your orders compare to your classmates'?

Problem-Solving Strategies
Use Logical Thinking

Sometimes a problem requires logical thinking. You must find an answer when you are missing some of the information. You must make inferences based on what you know. Often logic problems have many clues. Sometimes grids are used to organize the clues. Other times manipulatives are used.

Apple Arrays

Topic
Logical thinking

Key Question
How can you arrange four colors of apples in a square array so that each row, column, and small box has all four colors?

Learning Goal
Students will practice their logical thinking skills as they try to arrange apples in an array according to certain rules.

Guiding Document
NCTM Standards 2000*
- *Apply and adapt a variety of appropriate strategies to solve problems*
- *Build new mathematical knowledge through problem solving*

Math
Logic
Problem solving

Problem-Solving Strategies
Use logical thinking
Guess and check
Use manipulatives

Integrated Processes
Observing
Comparing and contrasting
Applying

Materials
Crayons or colored pencils
Scissors
Transparency film and pens
Student pages
Area Tiles, optional

Background Information
This activity provides an opportunity for students to apply deductive reasoning and logical thinking in an engaging format. The use of a manipulative allows them to guess and check without having a record of their incorrect attempts. As students arrange apples in an array, they will gain useful problem-solving skills that they can apply in other areas and contexts.

Management
1. Each student will need crayons or colored pencils in red, yellow, blue, and green.
2. Students may color and cut out the apple markers, or you may give them Area Tiles in the four colors to use instead.
3. To introduce the activity, make an additional copy of the first student page on transparency film. Cover the apples so you can fill the grids using different arrangements as examples.
4. If you do not have Area Tiles or a similar, colored manipulative for the overhead, color the apples with transparency pens and cut them apart.

Procedure
1. Put the transparency of the modified student page up on the overhead projector and partially fill the vertical grid as shown here using Area Tiles or the colored apples.
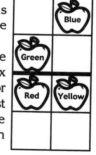
2. Explain the rules for putting the apples into the arrays. Each small box of four squares must have every color apple, and every row (or column) must have every color. This means that the same color will never appear twice in the same row, column, or small box.
3. Point to the bottom left square and ask students what color apple should go there and why. Have a volunteer come up and put an apple in the bottom left square and justify the choice. [A blue apple must go there. It cannot be red or yellow because red and yellow are already in the bottom small square. It cannot be green because green is already in the far left column.] If the choice is incorrect, ask the class if they agree and allow the student to make the necessary changes.

4. Repeat this process with the remaining squares, each time having a different student fill in the blank and justify his or her decision. If desired, do another practice problem with the horizontal grid.
5. Tell students that they will now do the same kind of problem on their own. Distribute the first student page, crayons, and scissors to each student. Have them color and cut out the apple markers at the bottom of the page.
6. Allow them to work in pairs or small groups to discover the solutions for the first two problems. Have them record their solutions by coloring in the arrays to correspond to the locations of the apples.
7. Once students are comfortable with the process, distribute the second student page. Explain that the challenge is the same, but the grids in this case are bigger, so it may take students more time to solve the problems. Remind them that every row, column, and small square must have all four colors of apples.
8. When students have discovered and recorded their solutions, close with a time of class discussion and sharing.

Connecting Learning
1. How did you decide where to put the first apple?
2. How did you know which apple belonged there?
3. Which problems were the easiest to solve? Why? Which problems were the hardest to solve? Why?
4. Was there ever more than one right answer? [no] Why or why not? [With four or more apples already in the grid, there is only one way to make all of the other apples fit according to the rules.]
5. Did it help to have the apples to move around? Would you have been able to solve the problems without the apples? Why or why not?

Extensions
1. Have students create their own problems and try them on classmates.
2. Give students an array that only has three apples filled in. Allow them to find as many different solutions as they can—there will be several possibilities.

Solutions

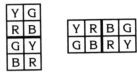

* Reprinted with permission from *Principles and Standards for School Mathematics*, 2000 by the National Council of Teachers of Mathematics. All rights reserved.

Apple Arrays

Learning Goal

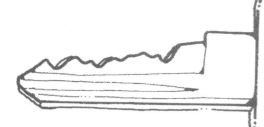

practice their logical thinking skills as the try to arrange apples in an array according to certain rules.

Apple Arrays

Fill the arrays with apples. All four colors must be in each row or column. All four colors must be in both small boxes.

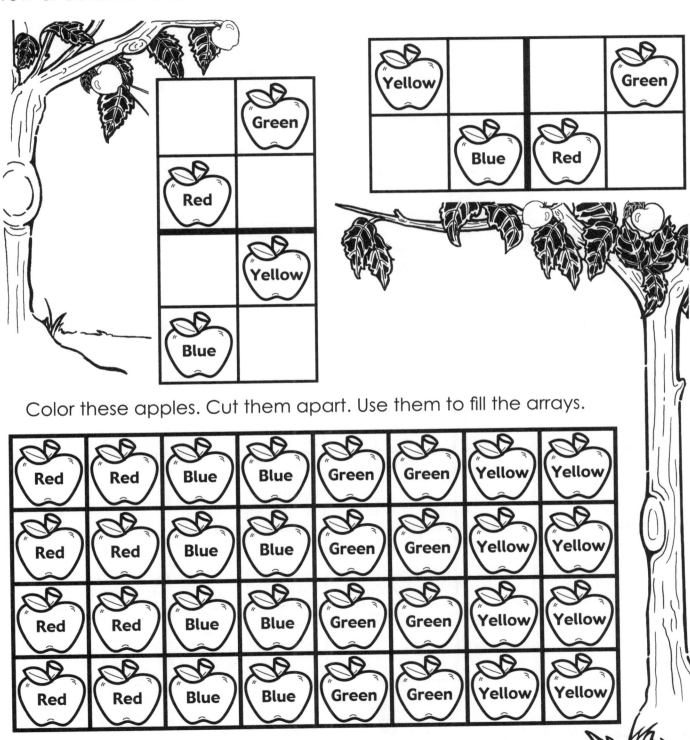

Color these apples. Cut them apart. Use them to fill the arrays.

Apple Arrays

Fill the arrays with apples. All four colors must be in each row and column.
All four colors must be in every small box.

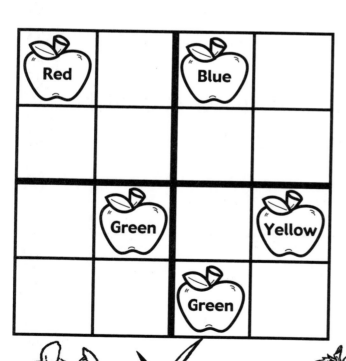

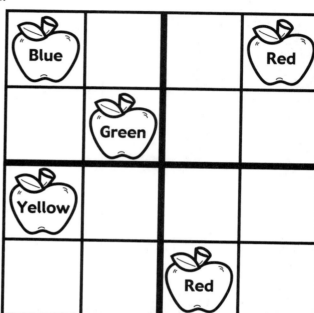

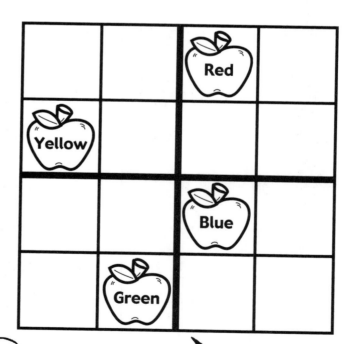

Apple Arrays

Connecting Learning

1. How did you decide where to put the first apple?

2. How did you know which apple belonged there?

3. Which problems were the easiest to solve? Why? Which problems were the hardest to solve? Why?

4. Was there ever more than one right answer? Why or why not?

5. Did it help to have the apples to move around? Would you have been able to solve the problems without the apples? Why or why not?

Historical *Logic*

Topic
Logical thinking

Key Question
How can you use the clues and logical thinking to solve a series of problems?

Learning Goal
Students will use logical thinking to solve a series of problems dealing with different historical topics.

Guiding Document
NCTM Standards 2000*
- *Build new mathematical knowledge through problem solving*
- *Solve problems that arise in mathematics and in other contexts*
- *Monitor and reflect on the process of mathematical problem solving*
- *Apply and adapt a variety of appropriate strategies to solve problems*

Math
Logic
Problem solving

Integrated Processes
Observing
Comparing and contrasting
Collecting and recording data
Interpreting data

Problem-Solving Strategies
Use logical thinking
Use manipulatives

Materials
Student pages
Manipulatives, optional

Background Information
Using logical thinking as a problem-solving strategy involves examining a set of clues and using deductive and inductive reasoning to explain why a solution is valid. Often in logic problems a grid is used to organize these clues. There are several ways to go about solving logic problems. It is important to expose the students to different strategies and allow them to choose the one that they are most comfortable with.

Some students find it helpful to use an X and checkmark system to keep track of the information gathered from each clue, while others prefer to use manipulatives on the grid to help them solve the logic problems. With the X and checkmark system, students read a clue and place an X in any box(es) that can be eliminated based on the clue. A checkmark is placed in any box that is correct based on the clue. When using manipulatives to solve logic problems, students place one manipulative in each space on the grid. Manipulatives are removed as the options they represent are eliminated, and those that remain identify correct answers. Whatever process is used, students should test their conclusions against the clues, making adjustments until they are satisfied with the solutions they have found.

Management
1. This activity contains a collection of logic problems relating to the history of the national parks, aviation, and American states. The problems have varying degrees of difficulty, so you will need to select the one(s) most appropriate for your students.
2. This activity can be presented as a whole-class activity, a small-group activity, or an individual activity. The various problems can be presented throughout the year as appropriate.
3. Provide manipulatives for those students who find them easier to work with than the X and checkmark system.

Procedure
1. Tell the students that they are going to solve several logic problems. Distribute the student page(s) and have students get into groups, if they will be working collaboratively.
2. Provide time for students to complete the problems.
3. After solving each puzzle, have students reread the clues to verify that the correct solution has been reached. This reinforces the importance of always checking your work, and gives students the opportunity to explain their reasoning, which develops their mathematical communication skills.
4. Close with a time of class discussion and sharing.

Connecting Learning
1. How did you solve the problems?
2. Which clues were the most helpful? Why?
3. What kind of clues were confusing?

4. What does a sentence or clue with the word *not* tell you? If *not* tells you that something is not true, how might it also tell you something that is true?
5. What kinds of words told you to place an X (or remove a manipulative)?
6. If a clue gave you a direct yes, what did that tell you about the boxes above, below, and beside the yes response?

Solutions

Which animal lives in which park?
The alligator lives in marshy swamps, and Death Valley is a big desert, so the alligator must live in the Everglades.
This means that the kangaroo rat lives in Death Valley.

Where are the parks located?
Crater Lake must be in Oregon, because the park with the lake is farthest north.
The Petrified Forest is not in New Mexico, and Crater Lake is in Oregon, so it must be in Arizona.
That means that Carlsbad Caverns is in New Mexico.

In what order were the parks created?
Sequoia must have been second because it has the shortest name.
Yellowstone must have been first because the oldest park begins with a "Y" and is not Yosemite.
That means Yosemite was third.

What plants live in the parks?
The giant sequoia grows in Kings Canyon.
Sea grass must live in Biscayne National Park because it is mostly ocean.
Saguaro National Park is a desert, so the cactus must live there.
This means that Alder trees live in Glacier National Park.

Where were these first flights?
The hot air balloon did not fly first in the United States, so it must have flown in France, which means that the airplane first flew in the United States.

When were these first flights?
The hot air balloon was first, so it must have been in 1783.
The glider was not in 1903, so it must have been in 1804.
This means the airplane first flew in 1903.

What were these "firsts" called?
The Graf Zeppelin must have been the first blimp to fly around the world since it did not carry mail or airplanes.

The Eagle Ovington must have been the first airplane to carry mail because it was not an aircraft carrier.
The USS Langley must have been the first aircraft carrier because it did not fly around the world, and the Eagle Ovington was the airplane.

Who was the first to make these flights?
Chuck Yeager was the first to fly faster than sound.
Ameilia Earhart was the first woman to fly solo across the Atlantic because she is the only woman in the group.
Cal Rodgers must have made the first flight across the United States because he flew in 1911 (so he could not have been the first to fly), and Chuck Yeager flew faster than sound.
This means that Orville Wright must have been the first to fly.

What are the state birds of these states?
Mississippi is on the Gulf of Mexico, so it must have the mockingbird as its state bird.
Hawaii is the island state, so it cannot have the robin and must have the nene.
This means that the robin is Michigan's state bird.

What is the order of the three smallest states in area from smallest to largest?
Rhode Island must be the smallest because Delaware is not, and Connecticut is larger than Rhode Island.
Deleware must be the 49th because it is not smallest or largest.
This means Connecticut must be the largest of the three.

What year did these states enter the Union?
Alaska must have entered in 1959 because that is 100 years after 1859.
Oregon must have entered in 1859 because it begins with "O," was not first, and entered before Oklahoma.
Virginia must have entered in 1788 because Oklahoma entered after Oregon.
This means that Oklahoma entered in 1907.

What is the order of the four largest states in population from largest to smallest?
California must be the largest because it is the farthest west.
Florida must be the fourth largest because it is the farthest south.
New York must be the second largest because it has more people than Texas.
This means Texas is the third largest.

*   Reprinted with permission from *Principles and Standards for School Mathematics*, 2000 by the National Council of Teachers of Mathematics. All rights reserved.

Historical *Logic*

Key Question

How can you use the clues and logical thinking to solve a series of problems?

Learning Goal

Students will:

use logical thinking to solve a series of problems dealing with different historical topics.

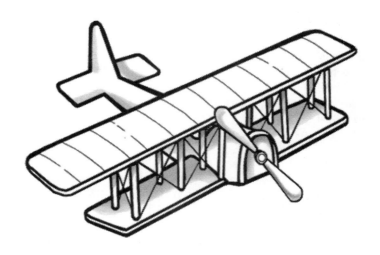

Historical Logic

Which animal lives in which park?

Clues:

- Death Valley is a big desert.

- The alligator lives in marshy swamp areas.

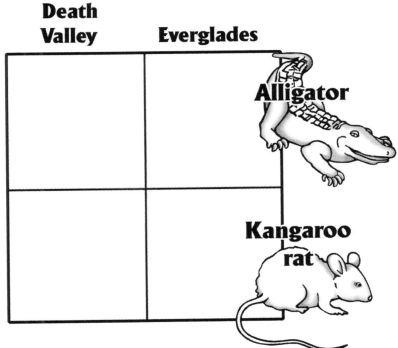

	Death Valley	Everglades

Where are the parks located?

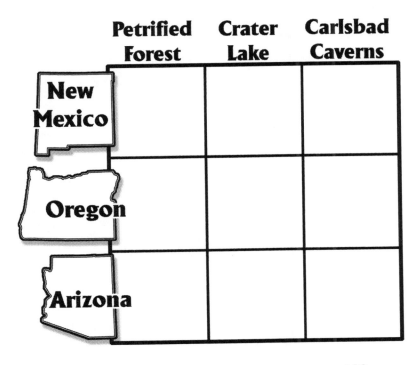

	Petrified Forest	Crater Lake	Carlsbad Caverns
New Mexico			
Oregon			
Arizona			

Clues:

- The park with the lake is the farthest north.

- The petrified forest is not in New Mexico.

- The place that has bats is not in Oregon.

Historical *Logic*

In what order were the parks created?

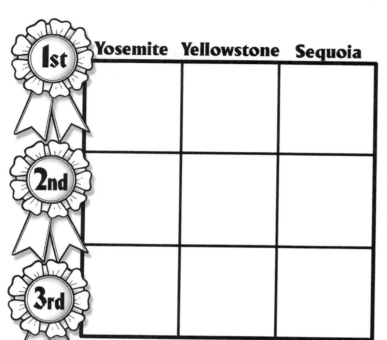

	Yosemite	Yellowstone	Sequoia
1st			
2nd			
3rd			

Clues:

- The park with the shortest name was second.

- The oldest park begins with a "Y."

- Yosemite is not the oldest park.

What plants live in the parks?

Clues:

- Alder trees can survive in cold weather and snow.

- Biscayne National Park is mostly ocean.

- Saguaro National Park is a desert.

- The giant sequoia grows in Kings Canyon.

	Saguaro	Biscayne	Kings Canyon	Glacier Bay	
					Sea Grass
					Giant Sequoia
					Cactus
					Alder

Historical Logic

Where were these first flights?

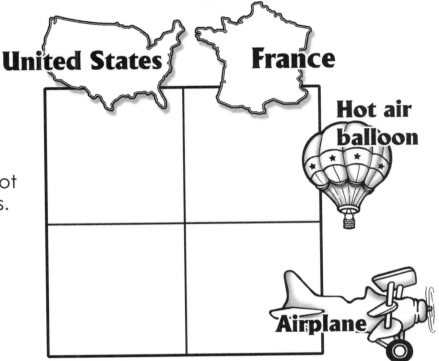

United States

France

Hot air balloon

Airplane

Clues:

- The hot air balloon was not flown in the United States.

When were these first flights?

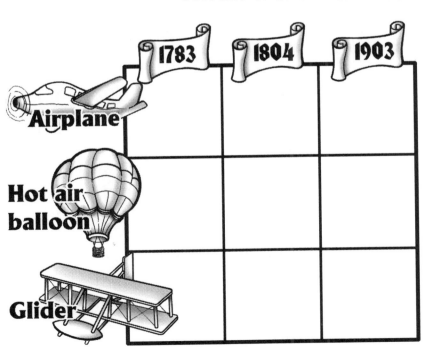

1783   1804   1903

Airplane

Hot air balloon

Glider

Clues:

- The hot air balloon was the first to be flown.

- The first glider was not flown in 1903.

Historical Logic

What were these "firsts" called?

Clues:

- The Eagle Ovington was not an aircraft carrier.

- The USS Langley did not fly around the world.

- The Graf Zeppelin did not carry mail or airplanes.

	Eagle Ovington	USS Langley	Graf Zeppelin
First aircraft carrier			
First blimp to fly around the world			
First airplane to carry mail			

Who was the first to make these flights?

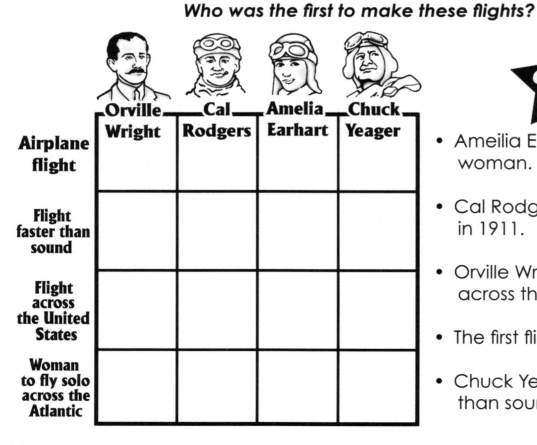

	Orville Wright	Cal Rodgers	Amelia Earhart	Chuck Yeager
Airplane flight				
Flight faster than sound				
Flight across the United States				
Woman to fly solo across the Atlantic				

Clues:

- Ameilia Earhart was a woman.

- Cal Rodgers made his flight in 1911.

- Orville Wright did not fly across the United States.

- The first flight was in 1903.

- Chuck Yeager flew faster than sound.

Historical *Logic*

What are the state birds of these states?

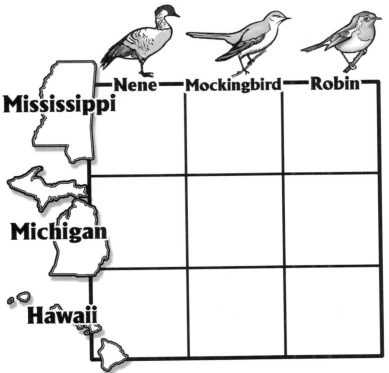

	Nene	Mockingbird	Robin
Mississippi			
Michigan			
Hawaii			

Clues:

- The island state does not have the robin as its bird.

- The bird with the longest name goes with the state on the Gulf of Mexico.

What is the order of the three smallest states in area from smallest to largest?

Clues:

- Delaware is not the smallest or the largest of the three.

- Connecticut is larger than Rhode Island.

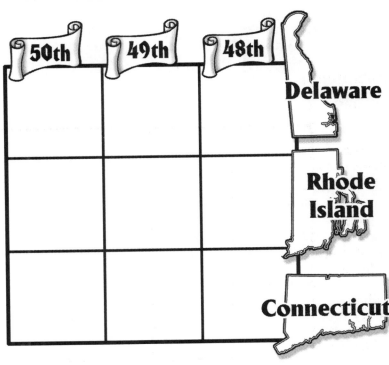

	50th	49th	48th
Delaware			
Rhode Island			
Connecticut			

Historical Logic

	1788	1859	1907	1959
Alaska				
Virginia				
Oregon				
Oklahoma				

Clues:

What year did these states enter the Union?

- Alaska entered exactly 100 years after a state that begins with "O."

- Oregon was not first to enter.

- Oklahoma entered after Oregon.

What is the order of the four largest states in population from largest to smallest?

Clues:

- The state farthest west has the most people.

- The state farthest north has more people than Texas.

- The state farthest south has the fourth largest population.

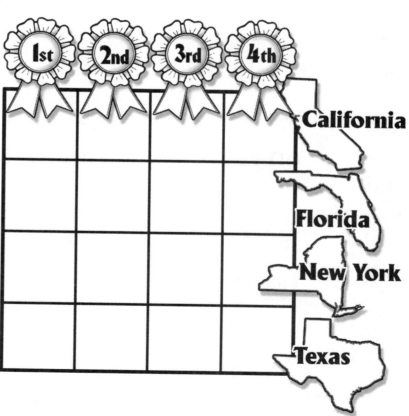

1st	2nd	3rd	4th	
				California
				Florida
				New York
				Texas

Connecting Learning

1. How did you solve the problems?

2. Which clues were the most helpful? Why?

3. What kind of clues were confusing?

4. What does a sentence or clue with the word *not* tell you? If *not* tells you that something is not true, how might it also tell you something that is true?

5. What kinds of words told you to place an X (or remove a manipulative)?

6. If a clue gave you a direct yes, what did that tell you about the boxes above, below and beside the yes response?

Topic
Logical thinking

Key Question
How can you work together using clues to determine what number you have?

Learning Goals
Students will:
1. work collaboratively in groups; and
2. use logic clues to determine what three-, four-, or five-digit number is being described.

Guiding Document
*NCTM Standards 2000**
- *Apply and adapt a variety of appropriate strategies to solve problems*
- *Build new mathematical knowledge through problem solving*
- *Develop understanding of the relative position and magnitude of whole numbers and of ordinal and cardinal numbers and their connections*

Math
Number sense
    odd and even
    place value
    greater than, less than
Logic
Problem solving

Integrated Processes
Observing
Comparing and contrasting
Collecting and recording data
Applying

Problem-Solving Strategies
Use logical thinking
Use manipulatives

Materials
Small counters (see *Management 2*)
Clue cards
Student page

Background Information
Logical thinking skills are important in many aspects of life besides mathematics. This activity provides an opportunity for students to sharpen their logical thinking while working collaboratively. Students must learn to listen carefully to the members of their group and work together to reach the correct solution. No one person can solve the logic problem alone. All four clues are needed to reach the answer.

This activity also provides an excellent opportunity for students to work with the kinds of vocabulary that they will encounter on standardized tests. They must be able to make a distinction between the entire number and a digit within that number, know odd from even, understand greater than/less than concepts, and be able to apply place value up to the ten thousands place.

Management
1. Students need to work in groups of four on this activity. If you do not have a number of students divisible by four, you can add a fifth student to some groups to be the official recorder.
2. Each group will need one copy of the student page and a set of 30-45 counters with which to cover up the numbers. These counters can be pennies, buttons, beans, area tiles, or any other small manipulative that will fit into the spaces provided.
3. This activity uses a lot of mathematical vocabulary. An understanding of odd and even, greater than/ less than, and place value is essential. If necessary, review terms and concepts before beginning.
4. There are eight sets of clue cards. Sets A and B deal with three-digit numbers; sets C, D, and E deal with four-digit numbers; and sets F, G, and H deal with five-digit numbers. Select the clue set that is appropriate for each group of students. (Multiple groups may use the same set, or each group may have its own.)
5. You will need to copy and cut apart the clue sets before beginning this activity. You may wish to laminate the clues and/or copy them onto card stock for repeated use.

Procedure

1. Have students get into their groups of four. Distribute one student page, one clue set, and an appropriate number of counters to each group.
2. Instruct each student in the group to take one clue card from the set. (If there are five students in the group, have the fifth student take the recording page and the counters.)
3. Inform students that the challenge is to work together as a group to figure out what number their clues describe. This is done by marking off numbers with the counters until only one possibility remains.
4. Explain that every student will take turns reading the clues on his or her card. If the clue gives information that can be immediately used, that information can be recorded by covering the numbers that are eliminated. (For example, if the clue says "There is no '7' in this number," all of the sevens on the page can be covered with a counter.) Once all of the clues have been read, they can share them as a group and revisit clues that may become more helpful once other information is known.
5. Give groups time to work on the clues. Encourage them to persevere until they come up with a solution and to check that solution against every clue to make sure that it is correct.
6. Have them justify their responses to you, or to the entire class, by going through each clue and showing how the answer meets the criteria.
7. Repeat this exercise throughout the year to continue to practice logic and critical vocabulary.

Connecting Learning

1. Were some clues more useful than others at first? Why or why not?
2. How did your group solve the problem?
3. How do you know that you have the correct answer?
4. What did you learn by doing this problem?

Extensions

1. Create your own logic clues for students, or allow them to create clues for each other.
2. For advanced students, go up to the hundred thousands place.

Solutions

A 153
B 809
C 1234
D 5391
E 2866
F 11,251
G 22,404
H 76,543

* Reprinted with permission from *Principles and Standards for School Mathematics*, 2000 by the National Council of Teachers of Mathematics. All rights reserved.

Clue Me In

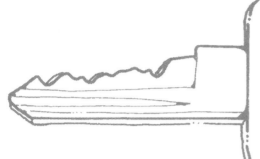

Key Question

How can you work together using clues to determine what number you have?

Learning Goals

Students will:

1. work collaboratively in groups; and
2. use logic clues to determine what three-, four-, or five-digit number is being described.

ten thousands	thousands	hundreds	tens	ones
0	0	0	0	0
1	1	1	1	1
2	2	2	2	2
3	3	3	3	3
4	4	4	4	4
5	5	5	5	5
6	6	6	6	6
7	7	7	7	7
8	8	8	8	8
9	9	9	9	9

Clue Me In

A

All of the digits in the number are odd.

The number is less than 900.

A

There is no "7" in the number.

The sum of the digits in the tens and hundreds places is six.

A

The digit in the ones place is less than five.

The sum of all three digits is nine.

A

The number in the ones place is two more than the number in the hundreds place.

The number is greater than 100.

Clue Me In

The digit in the ones place is one more than the digit in the hundreds place.

The number is less than 900.

The digit in the tens place is not odd.

The sum of the digits in the ones and tens places is nine.

The number is odd.

The digit in the hundreds place is even.

The number is greater than 700.

The sum of all the digits in the number is 17.

Clue Me In

The digit in the ones place is three more than the digit in the thousands place.

The number is even.

There are four digits in the number.

The number is less than 1500.

The digit in the hundreds place is one less than the digit in the tens place.

Half of the digits in the number are odd.

The digit in the tens place is "3."

No digits in the number are greater than five.

Clue Me In

D All of the digits in the number are odd.

The number is less than 6000.

D There is no "7" in the number.

The sum of the digits in the ones and tens places is 10.

D The number is more than 3000.

The number in the hundreds place is six less than the number in the tens place.

D When you add the digits in the thousands place and the hundreds place, you get eight.

Clue Me In

There are no odd digits in the number.

The number is less than 4000.

There is no "4" in the number.

The digits in the ones place and the tens place are the same.

The digit in the hundreds place is "8."

There are three different digits in the number.

The sum of the digits in the tens place and the thousands place is eight.

The number is more than 2500.

Clue Me In

There are only three different digits in this number.

There are four odd digits in this number.

The sum of all the digits in this number is 10.

The digit in the hundreds place is "2."

The digit in the thousands place is the same as the digit in the ones place.

The number is less than 12,000, but greater than 10,000.

There is no "3" in this number.

The digit in the tens place is three more than the digit in the hundreds place.

Clue Me In

The number is even.

The digit in the ones place is the same as the digit in the hundreds place.

The number has no odd digits.

The sum of the digits in the ten thousands place and the thousands place is less than five.

There are three different digits in this number.

The sum of the numbers in the tens place and the ones place is four.

The number is greater than 21,000.

The digit in the hundreds place is "4."

Clue Me In

The digit in the hundreds place is two less than the digit in the ten thousands place.

The digit in the tens place is even.

The sum of the digits in the ones and tens places is seven.

There is no "1" in this number.

The number in the thousands place is one more than the number in the hundreds place.

The number is greater than 60,000, but less than 90,000.

The number begins with an odd digit.

Every digit in this number is different.

Clue Me In

Connecting Learning

1. Were some clues more useful than others at first? Why or why not?

2. How did your group solve the problem?

3. How do you know that you have the correct answer?

4. What did you learn by doing this problem?

158

Problem-Solving Strategies
Work Backwards

Sometimes it's best to start at the finish when solving a problem. Working backwards helps when you know the answer, but don't know how to get there. You can start at the end and find the missing steps to get to the beginning.

What's the Question?

Topic
Problem solving

Key Question
If the answer is 10, what could the questions be?

Learning Goal
Students will work backwards to determine a variety of possible questions given an answer.

Guiding Document
NCTM Standards 2000*
- *Understand various meanings of addition and subtraction of whole numbers and the relationship between the two operations*
- *Understand the effects of multiplying and dividing whole numbers*
- *Express mathematical relationships using equations*
- *Build new mathematical knowledge through problem solving*

Math
Number and operations
Problem solving

Integrated Processes
Observing
Comparing and contrasting
Recording

Problem-Solving Strategies
Work backwards
Write a number sentence
Use manipulatives

Materials
Student page
Unifix cubes, optional

Background Information
This activity features an open-ended challenge that asks students to come up with multiple problems that produce a single answer. It draws a bit of inspiration from the popular TV game show *Jeopardy* in which contestants are given an answer and are challenged to come up with the correct question. Its main inspiration, however, comes from Arthur Wiebe and the late Larry Ecklund. These two gentlemen, the cofounders of AIMS, showed that powerful mathematics can happen in elementary classrooms when rich, open-ended mathematics problems are given to students in addition to the more mundane fare provided by most math textbooks.

This particular activity is a slight modification of an idea Larry Ecklund presented in a graduate math education class. He posed the rhetorical question that provides the main inspiration for this activity: "Which

is a richer mathematical experience for students, asking them the product of six and four or asking them to list all the whole number pairs whose product is 24?" This question presented a new way of viewing math teaching and showed how easily a typical drill and practice question could be converted into a powerful mathematical experience for students.

What's the Question? combines *Jeopardy's* reversal of the normal question-answer sequence with Larry Ecklund's rich multiple-answer questions. In this activity, students are given several numerical answers and then challenged to come up with as many questions as they can for each number given as an answer.

Management
1. The examples on the student page show the questions in both written and number sentence form. If appropriate, students should be challenged to do both. Writing the questions out in word form gives students practice using proper mathematical vocabulary.
2. If necessary, change the numbers for which students find the questions to be more appropriate for your class.
3. For students who may need a concrete representation, you may wish to provide two colors of Unifix cubes that they can put together in various combinations to make the target answer.

Procedure
1. Distribute the student page and go over the instructions with the class.
2. Provide time for students to complete the activity, making Unifix cubes available if desired.
3. Close with a time of class discussion and sharing.

Connecting Learning
1. What questions did you come up with for the answer 5? ...12? ...17? ...24? ...100?
2. How did these questions compare to the ones your classmates came up with?
3. Who is correct? [if the math is done properly, everyone]
4. What did you learn by doing this activity?

Extensions
1. Have students choose their own answers once they complete the ones listed on the student page.
2. Challenge students to include questions for each of the basic operations instead of using just one or two.

* Reprinted with permission from *Principles and Standards for School Mathematics*, 2000 by the National Council of Teachers of Mathematics. All rights reserved.

What's the Question?

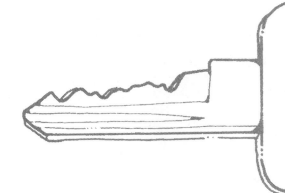

Key Question

If the answer is 10, what could the questions be?

Learning Goal

Students will:

work backwards to determine a variety of possible questions given an answer.

What's the Question?

The answer is 10.
Some possible questions that go with this answer include:
1. What is the sum of three and seven?   3 + 7 = ?
2. What is the sum of eight and two?   8 + 2 = ?
3. What is the product of two and five?   2 x 5 = ?
4. What is the difference between 23 and 13?   23 – 13 = ?
5. What is the quotient of 90 divided by 9?   90 ÷ 9 = ?

   In this activity, you start with the answer and work backwards. You must come up with questions and/or number sentences that fit the answer, as shown above. Try to come up with as many questions as you can for each of the following answers. List your questions in the space below. You may use the back of this page and additional paper if necessary.

The answers: 5, 12, 17, 24, and 100.

What's the **Question?**

Connecting Learning

1. What questions did you come up with for the answer 5? ...12? ...17? ...24? ...100?

2. How did these questions compare to the ones your classmates came up with?

3. Who is correct?

4. What did you learn by doing this activity?

What's the Scoop?

Topic
Problem solving

Key Question
How can you figure out what kinds of ice cream kids bought if you know how much money they spent?

Learning Goals
Students will:
1. work backwards to determine what kind of ice cream cone each person bought based on the amount of money each spent,
2. make their own orders by giving the correct amount of money, and
3. determine what their partners ordered based on the amount of money received.

Guiding Document
NCTM Standards 2000*
- *Develop a sense of whole numbers and represent and use them in flexible ways including relating, composing, and decomposing numbers*
- *Apply and adapt a variety of appropriate strategies to solve problems*
- *Solve problems that arise in mathematics and in other contexts*
- *Build new mathematical knowledge through problem solving*

Math
Number and operations
   addition
   subtraction
Problem solving

Integrated Processes
Observing
Recording
Applying

Problem-Solving Strategies
Work backwards
Use manipulatives
Write a number sentence

Materials
Part One:
   scissors
   glue sticks
   first four student pages

Part Two:
   play money (see *Management 2*)
   scissors
   tape
   final two student pages, laminated
     (see *Management 3*)

Background Information
   The strategy of working backwards is an important skill to have for situations when the answer, destination, outcome, etc., is known, but the steps used to get there are unknown. This strategy is useful in many different contexts, from shopping to driving to cooking. In this activity, students are given the opportunity to work backwards in the context of ordering treats from an ice cream shop. They will first gain practice by working out problems on their own and will then apply that experience by role playing with a partner.

Management
1. This activity is divided into two parts. In the first part, students gain experience by working on problems individually. In the second part, they apply what they have learned by working with a partner.
2. For *Part Two* of this activity, each pair of students will need enough play money (change and bills) so that they can order any combination of ice cream and toppings from the menu and pay for it with exact change.
3. Copy the pages of ice cream scoops and the cone onto card stock. Laminate one copy of each page for every pair of students to use in *Part Two*. If the pages are not laminated, students will be unable to use them repeatedly.

Procedure
Part One
1. Distribute the first four student pages, scissors, and a glue stick to each student. Have students cut out the page of menu items.
2. Go over the menu prices as a class. Do a couple of sample problems to make sure everyone understands how to calculate totals. For example, I am buying two scoops of ice cream with sprinkles. How much will that cost? Be sure that students understand that there will always be at least one topping, and that there may sometimes be two or three.

3. Explain that students will be trying to figure out what each person bought based on how much they spent. Once they have found a solution, they should glue down the appropriate pictures to make an equation.

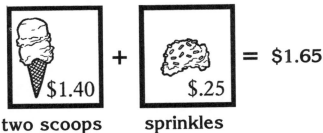

two scoops   sprinkles

4. Allow sufficient time for students to complete the student pages. Once everyone is done, discuss the solutions and how students went about solving the problems.

Part Two
1. Have students get into pairs and distribute the materials to each pair. Instruct them to cut out the pictures of the ice cream scoops. They should leave the cone intact in its box so they can tape scoops on it.
2. Explain that the students will take turns being the employee at the ice cream shop and a customer. Describe the responsibilities of each person.
   Customer: Decide what you want and count out the correct amount of money. Give your money to the employee and write down a number sentence describing what you ordered and how much it costs. Do not tell the employee what you are ordering.
   Employee: Determine what the customer wants based on the amount of money you are given. Tape the correct scoop(s) of ice cream onto the cone and give the customer his or her order. (Only the top scoop gets the toppings.)

3. Tell students that once the ice cream has been given, they should verify that the employee gave what the customer ordered. If there is a difference, determine which person was incorrect by looking at the number sentence and the amount of money paid.
4. Allow time for students to play each part several times and discuss the experience as a class.

Connecting Learning
1. How were you able to determine what each person ordered?
2. How did the number sentences help you check your answers?
3. Was it easier for you to be the customer or the worker? Why?
4. Did you get what you thought you were ordering every time? Why or why not?

Extensions
1. For *Part Two*, make your own menu with actual foods that students can order. For example, make-your-own trail mix—almonds $.70 a scoop, chocolate chips, $.50, raisins, $.35, etc.
2. Allow students to create their own menus and prices to practice with.

* Reprinted with permission from *Principles and Standards for School Mathematics*, 2000 by the National Council of Teachers of Mathematics. All rights reserved.

What's the Scoop?

Key Question

How can you figure out what kinds of ice cream kids bought if you know how much money they spent?

Learning Goals

Students will:

1. work backwards to determine what kind of ice cream cone each person bought based on the amount of money each spent,
2. make their own orders by giving the correct amount of money, and
3. determine what their partners ordered based on the amount of money received.

What's the Scoop?

What's the Scoop?

Show each solution. Use your pictures to make a number sentence.

1. Tamara spent $1.20. What did she get?

= $1.20

2. Diego spent $1.75. What did he get?

= $1.75

3. Audra spent $1.55. What did she get?

= $1.55

4. Hassan spent $2.15. What did he get?

= $2.15

168

5. Mikela spent $1.90. What did she get?

= $1.90

6. Jacob spent $2.70. What did he get?

= $2.70

7. Jamillah spent $2.50. What did she get?

= $2.50

8. Daniel spent $1.80. What did he get?

= $1.80

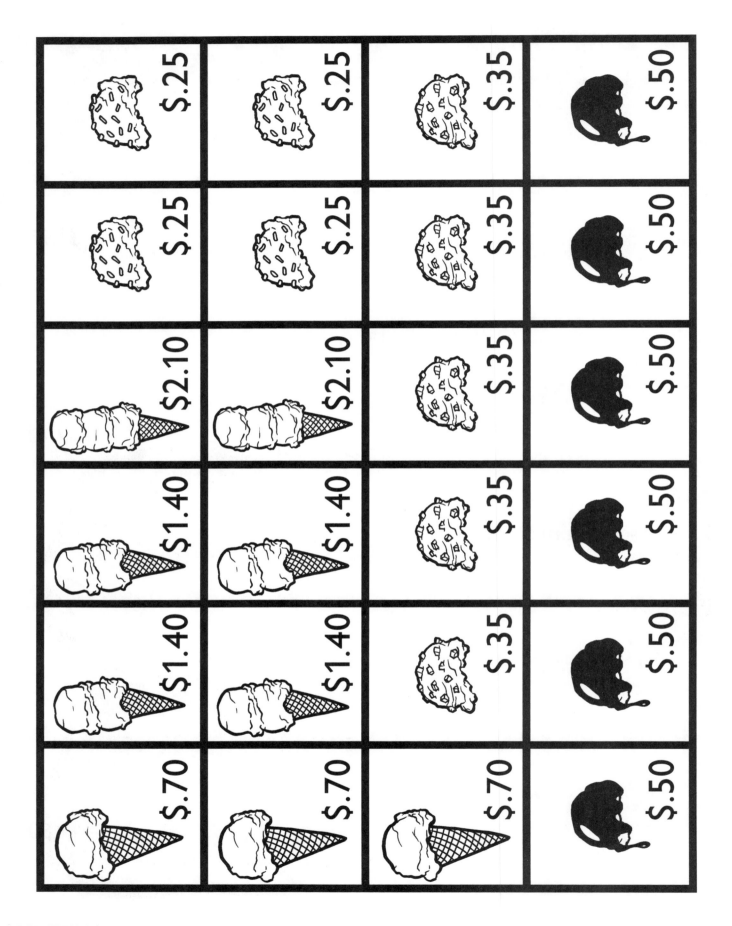

What's the Scoop?

Connecting Learning

1. How were you able to determine what each person ordered?

2. How did the number sentences help you check your answers?

3. Was it easier for you to be the customer or the worker? Why?

4. Did you get what you thought you were ordering every time? Why or why not?

Time Tellers

Topic
Elapsed time

Key Question
How can we decide when to begin a series of tasks if we know how long they will take and when they must be finished?

Learning Goals
Students will:
1. work backwards to determine the appropriate starting time for the day's activities,
2. use a bus and movie schedule to determine when they need to leave home in order to see a specific movie, and
3. determine which attractions can be seen on a field trip to the zoo.

Guiding Document
*NCTM Standards 2000**
- *Recognize the attributes of length, volume, weight, area, and time*
- *Build new mathematical knowledge through problem solving*
- *Solve problems that arise in mathematics and in other contexts*
- *Apply and adapt a variety of appropriate strategies to solve problems*

Math
Measurement
    time
Problem solving

Integrated Processes
Observing
Comparing and contrasting
Applying
Relating

Problem-Solving Strategy
Work backwards

Materials
Student mini books
Scratch paper

Background Information
Sometimes the only way to tackle a problem is to look at the final result and move backwards to the beginning through a series of steps. Often this type of problem asks a question about how much one started with or when something began or should begin. Because the problem itself asks a question about the beginning rather than the ending of the problem, it gives us the idea to work backwards with what we are given in order to arrive at that beginning value. This working backwards problem-solving strategy is often useful in planning an event or trying to determine quantities of materials needed for a particular project.

In *Part One* of this activity, students will become comfortable with the process of working backwards for time management purposes. In *Part Two* and *Part Three*, they will be introduced to real-world situations that will require them to use schedules to solve problems.

Management
1. Prior to this lesson, make copies of the student mini books—one per student. To make a mini book, fold the page in half horizontally and vertically so that the pages go in order.
2. This activity is not intended as an introduction to elapsed time. It is assumed that students have had prior experience dealing with elapsed time.

Procedure
Part One—Time Tellers
1. Ask the *Key Question*. Discuss the reasonableness of the responses.
2. Give each student a copy of the *Time Tellers* mini book.
3. Allow time for each student to read the story and arrive at an answer.
4. Have students share their answers with the people sitting around them. If there are differences, have each group discuss the processes they went through to determine the starting time.
5. Ask the groups to share their solutions with the class. If there are differences, decide as a class which answer is correct.

Part Two—Time Tables

1. Ask the students to brainstorm situations when they have worked backwards in order to be on time for certain events. If it is not mentioned, suggest the use of schedules for buses, trains, and subways.
2. Tell the class that you have another mini book for them and that this time they will be solving problems that will require them to use a movie and bus schedule.
3. Distribute the *Times Tables* mini book and discuss the problem.
4. Allow time for the students to read the story and answer the questions.
5. Discuss their answers and encourage students to share their strategies.

Part Three—A Trip to the Zoo

1. Ask the students if they have ever gone on a field trip to a place that had scheduled events, such as a science museum, where they offer different sessions at different times. Discuss how they have to decide which events to go to based on when the events start and how much time they have to be there.
2. Tell the class that they are going to solve several problems based on a field trip to the zoo and the scheduled events at the zoo.
3. Distribute the *A Trip to the Zoo* mini book and discuss the problems they will be expected to solve.
4. Provide time for the students to read the mini book and solve the problems.
5. Discuss the students' answers and strategies they used to arrive at the answers.

Connecting Learning

1. How did you decide when Nick should start his day?
2. Would you have been able to solve any of the problems without working backwards? Why or why not?
3. If it is 3:00 and your favorite television show comes on at 5:00, how long will you need to wait to see it? [two hours]
4. How did you decide what time Monique should leave home?
5. Why is it important to know how long an event takes and how much time you have?

Extension

Have students write their own elapsed time stories using *Time Tellers* as an example.

* Reprinted with permission from *Principles and Standards for School Mathematics*, 2000 by the National Council of Teachers of Mathematics. All rights reserved.

Time Tellers

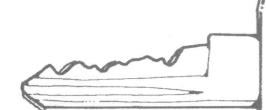

Key Question

How can we decide when to begin a series of tasks if we know how long they will take and when they must be finished?

Learning Goals

Students will:

1. work backwards to determine the appropriate starting time for the day's activities,

2. use a bus and movie schedule to determine when they need to leave home in order to see a specific movie, and

3. determine which attractions can be seen on a field trip to the zoo.

He wants to play video games with his brother Andrew. He knows that always takes an hour and a half.

He wants to go fishing with Jalisa and Andrew. It takes 15 minutes to walk to their favorite fishing place, and they usually fish for two hours.

The last thing he wants to do before he plays basketball is go skateboarding. It takes him 45 minutes to go to the park, ride his board, and get back.

What time does Nick need to start his day so that he will be ready to play basketball at 5:30?

Time Tellers

Nick promised to meet Miguel to play basketball at the school at 5:30 P.M. It is Saturday and he has other things to do before he plays basketball.

Movie Schedule

Action Film		
2:00	3:00	4:00

Comedy Film		
2:15	4:15	5:15

Family Film		
2:45	3:30	5:00

Bus Schedule

Blue Line	Red Line	Green Line
1:30	1:45	3:00
2:30	2:45	4:00
3:30	4:45	5:00

When should Monique leave home for the second showing of the action film?

Which bus will she take?

When should she leave home for the last showing of the comedy?

Which bus will she take?

If Monique leaves her house at 2:20, which bus will she most likely take, and which movie is she most likely going to see?

Time Tables

Monique wants to go to the movies. It takes 15 minutes by bus to get to the movie theater. The bus stop is 10 minutes from her house.

SOLVE IT! 3rd

Special Attractions

Banquet with the Birds	10:00	12:00 2:00
Camel Rides	11:30 12:30	1:30 2:30
Seal Feeding	10:30 11:30	12:30 1:30

Shows

Winged Wonders	10:00 11:00	12:00 1:00
Primate Playtime	10:00 11:00	12:00 1:00
Reptile Show	12:00 1:00	3:00 4:00

If each show and attraction is 45 minutes long, how many can Juan see during his visit?

What shows and attractions are they?

Can Juan have a camel ride, see the winged wonders show and the reptile show?

Why or why not?

A Trip to the ZOO

Juan's class took a field trip to the zoo. They arrived at 10:00 and have to meet back at the bus at 2:00.

Time Tellers

Connecting Learning

1. How did you decide when Nick should start his day?

2. Would you have been able to solve any of the problems without working backwards? Why or why not?

3. If it is 3:00 and your favorite television show comes on at 5:00, how long will you need to wait to see it?

4. How did you decide when Monique should leave home?

5. Why is it important to know how long an event takes and how much time you have?

Topic
Problem solving

Key Question
How can you use pattern blocks as manipulatives to solve fraction problems?

Learning Goal
Students will use pattern blocks to solve problems involving whole numbers, unit fractions, and mixed numbers.

Guiding Documents
Project 2061 Benchmark
- *Many objects can be described in terms of simple plane figures and solids. Shapes can be compared in terms of concepts such as parallel and perpendicular, congruence and similarity, and symmetry. Symmetry can be found by reflection, turns, or slides.*

NCTM Standards 2000*
- *Develop understanding of fractions as parts of unit wholes, as parts of a collection, as locations on number lines, and as divisions of whole numbers*
- *Build new mathematical knowledge through problem solving*

Math
Fractions
    mixed numbers
Geometry
    shapes
Problem solving

Integrated Processes
Classifying
Comparing and contrasting

Problem-Solving Strategies
Work backwards
Use manipulatives

Materials
Pattern blocks (see *Management 1*)
Student pages
Overhead transparency of student pages, optional

Background Information
One of the big ideas in fractions is the concept of "What is the one?" The "one" is called a "unit whole" in more precise mathematical terms. Once students realize that the one can change and that fractions represent equal parts of any given one, they are well on their way to building an understanding of fractions. Pattern blocks are a good manipulative to help students build fraction concepts. For example, if the hexagon is the one, the trapezoid is one-half, the rhombus is one-third, and the equilateral triangle is one-sixth. However, if the trapezoid is one, the hexagon is two, the rhombus is two-thirds, and the equilateral triangle is one-third.

While the focus of this lesson is not geometry, the pattern blocks are plane figures. The yellow piece is a regular hexagon, the red piece is a trapezoid, the blue piece is a rhombus or parallelogram, and the green piece is an equilateral triangle.

Management
1. Students will need access to pattern blocks for this activity. One tub is enough for a whole class since each student needs only one hexagon, two trapezoids, three rhombuses, and six equilateral triangles.
2. This should not be students' first exposure to basic fraction concepts.
3. This activity can be done by students working alone, in groups, or as a whole-class activity.
4. Overhead transparencies of the student pages can help facilitate the activity introduction and the sharing session at the end of the lesson.
5. While precise geometric vocabulary is not the focus of this lesson, it can be easily incorporated by using the correct terminology for the pattern block pieces.

Procedure
1. Distribute the pattern blocks and student pages.
2. Introduce the activity, making sure students understand what they are to do.

3. Facilitate students' work on the problems.
4. Lead a whole-class discussion at the end of the activity.

Connecting Learning

1. If the equilateral triangle is one, what whole number is represented by the rhombus? [two]
2. If the equilateral triangle is one, what whole number is represented by the hexagon? [six]
3. If the equilateral triangle is one, what whole number is represented by the trapezoid? [three]
4. If the rhombus is one, what whole number is represented by the hexagon? [three]
5. If the trapezoid is one, what whole number is represented by the hexagon? [two]
6. If the hexagon is one, what fraction is represented by the trapezoid? [one-half]
7. If the hexagon is one, what fraction is represented by the equilateral triangle? [one-sixth]
8. If the hexagon is one, what fraction is represented by the rhombus? [one-third]
9. If the trapezoid is one, what fraction is represented by the equilateral triangle? [one-third]
10. If the rhombus is one, what fraction is represented by the equilateral triangle? [one-half]
11. If the hexagon is one, what fraction is represented by two equilateral triangles? [two-sixths or one-third]
12. If the trapezoid is one, what fraction is represented by two equilateral triangles? [two-thirds]
13. If the hexagon is one, what fraction is represented by an equilateral triangle and a rhombus? [one-half or three-sixths]
14. If the hexagon is one, what fraction is represented by an equilateral triangle and a trapezoid? [four-sixths or two-thirds]
15. If the rhombus is one, what mixed number is represented by the trapezoid? [one and one-half]
16. If the trapezoid is one, what fraction is represented by the rhombus? [two-thirds]

Extensions

1. Have students use pattern blocks to explore mixed numbers in more depth, choosing various shapes or combinations of shapes as the "one."
2. Have students make up their own pattern block fraction problems and exchange them with other students.

* Reprinted with permission from *Principles and Standards for School Mathematics*, 2000 by the National Council of Teachers of Mathematics. All rights reserved.

BLOCKING OUT FRACTIONS

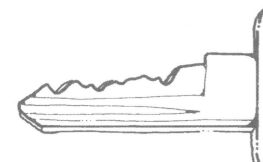

Key Question

How can you use pattern blocks as manipulatives to solve fraction problems?

Learning Goal

Students will:

use pattern blocks to solve problems involving whole numbers, unit fractions, and mixed numbers.

BLOCKING OUT FRACTIONS

Use pattern block pieces to answer the following questions.

1. If the is one, what whole number is the  ?

2. If the △ is one, what whole number is the ⬡ ?

3. If the △ is one, what whole number is the ▱ ?

4. If the ▱ is one, what whole number is the ?

5. If the ◸ is one, what whole number is the ⬡ ?

BLOCKING OUT FRACTIONS

Use pattern block pieces to answer the following questions.

6. If the ⬡ is one, what fraction is the ⬠ ?

7. If the ⬡ is one, what fraction is the △ ?

8. If the ⬡ is one, what fraction is the ▱ ?

9. If the ⬠ is one, what fraction is the △ ?

10. If the ▱ is one, what fraction is the △ ?

BLOCKING OUT FRACTIONS

Use pattern block pieces to answer the following questions.

11. If the ⬡ is one, what fraction is △ △ ?

12. If the ⬡ is one, what fraction is △ △ ?

13. If the ⬡ is one, what fraction is △ ◇ ?

14. If the ⬡ is one, what fraction is △ ⬜ ?

15. If the ◇ is one, what mixed number is the ⬜ ?

16. If the ⬜ is one, what fraction is the ◇ ?

Connecting Learning

1. If the equilateral triangle is one, what whole number is represented by the rhombus?

2. If the equilateral triangle is one, what whole number is the hexagon?

3. If the equilateral triangle is one, what whole number is represented by the trapezoid?

4. If the rhombus is one, what whole number is represented by the hexagon?

5. If the trapezoid is one, what whole number is represented by the hexagon?

6. If the hexagon is one, what fraction is represented by the trapezoid?

7. If the hexagon is one, what fraction is represented by the equilateral triangle?

8. If the hexagon is one, what fraction is represented by the rhombus?

Connecting Learning

9. If the trapezoid is one, what fraction is represented by the equilateral triangle?

10. If the rhombus is one, what fraction is represented by the equilateral triangle?

11. If the hexagon is one, what fraction is represented by two equilateral triangles?

12. If the trapezoid is one, what fraction is represented by two equilateral triangles?

13. If the hexagon is one, what fraction is represented by an equilateral triangle and a rhombus?

14. If the hexagon is one, what fraction is represented by an equilateral triangle and a trapezoid?

15. If the rhombus is one, what mixed number is represented by the trapezoid?

16. If the trapezoid is one, what fraction is represented by the rhombus?

Problem-Solving Strategies
Wish for an Easier Problem

Sometimes a problem has lots of data or big numbers. It can seem too hard to do. This is when you can "wish for an easier problem." You can use smaller numbers instead of the big ones. You can think about how to solve the problem instead of the numbers and data. This will help you see how to solve a simpler version of a harder problem.

Tallying Toothpick Tri▲ngles

Topic
Functions

Key Question
How many toothpicks are needed to build any number of equilateral triangles?

Learning Goals
Students will:
1. determine the number of toothpicks it takes to build one through five equilateral triangles, and
2. use this information to generalize a rule for how many toothpicks it takes to make any number of triangles.

Guiding Documents
Project 2061 Benchmark
- *Mathematics is the study of many kinds of patterns, including numbers and shapes and operations on them. Sometimes patterns are studied because they help to explain how the world works or how to solve practical problems, sometimes because they are interesting in themselves.*

NCTM Standards 2000*
- *Describe, extend, and make generalizations about geometric and numeric patterns*
- *Represent and analyze patterns and functions, using words, tables, and graphs*
- *Model problem situations with objects and use representations such as graphs, tables, and equations to draw conclusions*
- *Investigate how a change in one variable relates to a change in a second variable*
- *Build new mathematical knowledge through problem solving*

Math
Algebra
   functions
Patterns
Problem solving

Integrated Processes
Observing
Collecting and recording data
Analyzing data
Generalizing
Applying

Problem-Solving Strategies
Wish for an easier problem
Look for patterns
Organize the information
Use manipulatives

Materials
Flat toothpicks, 15 per student
Student page

Background Information
This activity is intended to help students construct a deeper understanding of the function concept. This key idea has its roots in the patterns with which primary students love to work. Unfortunately, the vital connection between patterns and functions is not articulated as often as it might be. Let's look at how these two concepts are linked so that we might better understand this important connection.

The big idea behind patterns is that they are predictable, and thus can be extended indefinitely. This predictability leads to a rule or generalization. If this rule produces a single, unique output for each input, it describes a special mathematical relationship called a function.

To illustrate the progression from pattern to rule to function, let's consider the following example. Suppose primary students are given a sheet of paper showing a row of 10 boxes with the first four boxes colored red, blue, red, blue. They will have no problems when asked to extend this pattern by coloring in the next six boxes. Since this pattern is very predictable, students can come up with a rule or generalization for it. One such rule is that the first box and every other box after that is colored red and that the second box and every other box after that is colored blue. Another way to state this generalization is to say that the odd-numbered boxes are red and the even-numbered boxes are blue.

Up to this point in our example, students have studied a pattern, come up with a generalization for that pattern, and applied this generalization to extend the pattern. While this is all that younger students should be asked (or expected) to do, older students are ready for the next step—studying the pattern as a function. To do this requires looking at the special relationship between two different sets of things, inputs and outputs. Perhaps the easiest way to see this relationship is to construct a table. We'll start by making a table for the red boxes.

The top row lists the inputs—in this case, the ordinal numbers (first, second, third, etc.) representing each red box. The bottom row shows the outputs—in this case the odd numbers—that correspond to the positions of the boxes.

red box number (input)	1	2	3	4	5	n
red box position (output)	1	3	5	7	9	$2n - 1$

When each input in the table above is paired with its corresponding output, a functional relationship between these two sets of numbers is established.

One of the ways this relationship can be shown or described is to use ordered pairs. For example, the first red box is in position one. This can be represented as the ordered pair (1, 1). Likewise, the second red box is in position three (2, 3), the third in position five (3, 5), and so on for each ordered pair. In each case, exactly one output is paired with each input, making this a functional relationship.

Another way to show this function is with an equation that produces a single output for each input. In our example, the equation linking the red boxes and their positions is derived from the generalization for the odd numbers ($2n - 1$). This equation can be written as $p = 2n - 1$, where p represents the position of a red box and n represents the ordinal number of any red box.

To describe this special relationship between inputs and outputs in the correct mathematical language, you would say that the position of any red box is a function of its number (first, second, etc.) in the sequence.

The one common element in each of these ways of showing or describing the function is the special connection between the inputs and the outputs in which every input produces exactly one predictable output. Learning this is essential to constructing a better understanding of functions.

Management

1. This activity is divided into two parts. In the first part, students are challenged to find the number of toothpicks needed to build 47, 365, or any number of triangles. In the second part, students are challenged to show and describe the mathematical relationship they discovered in the first part in as many ways as they can. This part can be done individually, in groups, or as a whole-class exercise depending on your students' needs and levels of ability.
2. Each student will need his or her own copy of the student page and at least 15 toothpicks. Flat toothpicks work best because they do not roll.

Procedure

1. Distribute the student page and toothpicks to each student.
2. State the *Key Question* and be sure that everyone understands the task.
3. Provide time for students to complete *Part One* of the activity.
4. If desired, have students get into groups to complete *Part Two*.
5. Conduct a time of class discussion and sharing where groups tell about the methods they came up with to describe the functional relationship. (See *Solutions* for examples of methods.)

Connecting Learning

1. How many toothpicks does it take to make one triangle? [three] ...two triangles? [six] ...five triangles? [15]
2. What pattern do you notice? [The number of toothpicks is always three times the number of triangles.]
3. How many toothpicks would it take to make 47 triangles? [141] ...365 triangles? [1095] How do you know without building them?
4. What are some of the ways your group came up with to describe the relationship between the number of toothpicks and the number of triangles? (See *Solutions*.)

Solutions

The following section details some of the ways students might show or describe the functional relationship between the triangles and toothpicks. This list is not intended to be comprehensive. You and your students may find additional ways to show or describe the function.

1. An arithmetic sequence
   While it only describes part of the functional relationship, one of the first things students should discover in this activity is the number sequence that results from their tally of the toothpicks used to build the triangles. This number sequence of 3, 6, 9, 12, 15, ... is easily extended by adding three to each previous value. A rule or generalization for going from one term in the sequence to the next can be summarized as *plus three*.
2. A simple rule
   This functional relationship can be stated as a simple rule. Some examples follow.
   • For each triangle, you need three toothpicks.
   • You need three times as many toothpicks as the number of triangles you want to build.
   • You need three toothpicks for every triangle.
3. A generalization or equation
   Since each triangle takes three toothpicks, students can describe this relationship as a generalization or equation. A few examples follow.

- Toothpicks equal three times the triangles.
- If T represents the number of triangles and t represents the number of toothpicks, then $3T = t$ and $t/3 = T$.
- If x represents the number of triangles and y represents the number of toothpicks, then $y = 3x$.

4. A table
The relationship can be shown using the following table.

Number of triangles	0	1	2	3	4	5	6	n
Number of toothpicks	0	3	6	9	12	15	18	$3n$

5. A set of ordered pairs
The information in the above table can be used to build a set of ordered pairs. In each ordered pair, the first number represents the triangles and the second the toothpicks in this manner, (triangles, toothpicks). The set of ordered pairs can be shown as follows: {(0, 0), (1, 3), (2, 6), (3, 9), (4, 12), (5, 15), ...}

6. A function machine
Many math texts introduce the function concept with a function machine like the one pictured below. This machine takes an input, processes it according to a rule, and produces a single output. In this case the number of triangles wanted is input, the machine multiplies this number by three, and the number of toothpicks needed is output.

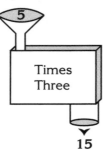

7. Formal statement
There is a functional relationship between the number of toothpicks and the number of triangles. This formal mathematical relationship can be stated as a sentence: The number of toothpicks needed is a function of the number of triangles built. While this may not seem very profound, it is a key understanding that needs to be built before students encounter such statements as "y is a function of x" or "$f(x)$ is a function of x" later on in their school careers.

* Reprinted with permission from *Principles and Standards for School Mathematics*, 2000 by the National Council of Teachers of Mathematics. All rights reserved.

Tallying Toothpick Tri△ngles

Learning Goals

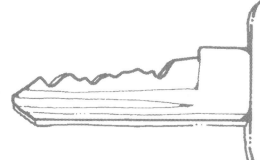

Students will:

1. determine the number of toothpicks it takes to build one through five equilateral triangles, and
2. use this information to generalize a rule for how many toothpicks it takes to make any number of triangles.

Tallying Toothpick Tri△ngles

Part One

How many toothpicks would you need to build 47 equilateral triangles? What about 365 triangles? Can you find a rule that will work for any number of triangles?

To answer these questions, use flat toothpicks to build five triangles. Tally the total number of toothpicks used for the first triangle and each additional triangle. Try to find a rule that describes how the number of toothpicks relates to the number of triangles. (This relationship is called a function.) Use your rule to answer the above questions. Show your work below.

Part Two

Describe the relationship between the triangles and toothpicks in as many ways as you can. Share your methods with the class.

Tallying Toothpick Tri△ngles

Connecting Learning

1. How many toothpicks does it take to make one triangle? … two triangles? …five triangles?

2. What pattern do you notice?

3. How many toothpicks would it take to make 47 triangles? …365 triangles? How do you know without building them?

4. What are some of the ways your group came up with to describe the relationship between the number of toothpicks and the number of triangles?

One Step at a Time

Topic
Problem solving

Key Question
How many steps would you take in 100 miles?

Learning Goals
Students will:
1. determine the approximate number of their steps in a portion of a mile, and
2. use what they have learned to determine how many of their steps would be in 100 miles.

Guiding Document
*NCTM Standards 2000**
- *Carry out simple unit conversions, such as from centimeters to meters, within a system of measurement*
- *Select and apply appropriate standard units and tools to measure length, area, volume, weight, time, temperature, and the size of angles*
- *Apply and adapt a variety of appropriate strategies to solve problems*
- *Build new mathematical knowledge through problem solving*

Math
Measurement
    linear
Problem solving

Integrated Processes
Observing
Collecting and recording data
Organizing data
Applying

Problem-Solving Strategies
Wish for an easier problem
Organize the information

Materials
Student page
Measuring tools (see *Management 2*)

Background Information
When a problem has large numbers and is overwhelming, we can "wish for an easier problem," or simplify the problem by solving a similar, easier problem. This activity can be easily solved if students focus on how many of their steps are in a half-mile, one-tenth of a mile, etc., rather than how many are in

100 miles. This makes the problem less complex and more manageable. They can multiply their answers to the simpler problems by the appropriate values to determine how many of their steps are in 100 miles.

Management
1. Students should work together in small groups of no more than three students.
2. Each group will need measuring tools such as a yardstick, ruler, etc., as well as a source where they can find the number of feet in a mile, feet in a yard, etc. These things should be available upon request. Do not suggest these to the students, but allow them to decide what they need to solve the problem.
3. Students should be allowed to determine what constitutes a step. For example, measuring from heel to heel, heel to toe, toe to toe, etc.

Procedure
1. Ask students the *Key Question.*
2. Have students get in groups and distribute a student page to each student.
3. Ask students what ideas they have about how they could solve the problem.
4. Provide time for students to work on the problem and record their methods and solutions.
5. Close with a time of sharing in which students explain how they arrived at their solutions.

Connecting Learning
1. What did you learn by starting with an easier problem?
2. How did what you learned help you solve the more complex problem?
3. How many of your steps are in 100 miles?
4. How does this compare to your classmates' numbers of steps?
5. How do you think your value would compare to your teacher's number of steps? Why?

Extension
Have students solve the extra challenge and use what they learned to find the number of their steps in 1000 miles.

* Reprinted with permission from *Principles and Standards for School Mathematics*, 2000 by the National Council of Teachers of Mathematics. All rights reserved.

One Step at a Time

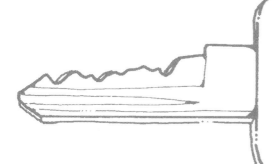

Key Question

How many steps would you take in 100 miles?

Learning Goals

Students will:

1. determine the approximate number of their steps in a portion of a mile, and
2. use what they have learned to determine how many steps are in 100 miles.

One Step at a Time

In colonial times, some people traveled by walking. They often had to walk between places as far apart as 100 miles.

Challenge: How many steps would you take in a 100-mile trip?

What would be an easier problem?

How are you going to answer the easier problem?

How can you use the information from the easier problem to find the number of your steps in 100 miles?

How many of your steps are in 100 miles?

Extra Challenge: Find the number of your steps in 1000 miles.

Leaving Boston
Philadelphia:
763,089 steps

One Step at a Time

Connecting Learning

1. What did you learn by starting with an easier problem?

2. How did what you learned help you solve the more complex problem?

3. How many of your steps are in 100 miles?

4. How does this compare to your classmates' numbers of steps?

5. How do you think your value would compare to your teacher's number of steps? Why?

Cube Face Estimation

Topic
Problem solving

Key Question
How many faces are on the cubes in the estimation jar?

Learning Goal
Students will estimate the number of faces on the cubes in the estimation jar.

Guiding Document
*NCTM Standards 2000**
- *Use geometric models to solve problems in other areas of mathematics, such as number and measurement*
- *Recognize geometric ideas and relationships and apply them to other disciplines and to problems that arise in the classroom or in everyday life.*
- *Apply and adapt a variety of appropriate strategies to solve problems*
- *Monitor and reflect on the process of mathematical problem solving*

Math
Estimation
Geometry
    properties of 3-D shapes
Number and operations
    multiplication or repeated addition
Problem solving

Integrated Processes
Observing
Applying

Problem-Solving Strategy
Wish for an easier problem

Materials
Wooden cubes, 30 or more
Large jar
Student page

Background Information
This activity provides an intriguing integration of geometry and number sense. Students are asked to determine the number of cube faces in a jar of cubes. If they try to count all the faces, the task is quite daunting. However, by breaking the problem down into smaller chunks, students can more easily make a good estimation. The first task will be to estimate the number of cubes. They then need to realize that each cube has six faces. By repeated addition or multiplication, students can arrive at an estimation.

Management
1. Make an estimation jar and fill it with at least 30 wooden cubes. Large gallon-size plastic jars work well. If these are not available, clear plastic shoeboxes or other clear containers will work.
2. As with all problem-solving situations, the concluding discussion is crucial. Students will learn many strategies from their classmates.

Procedure
1. Arrange students in small groups of two or three.
2. Ask the *Key Question* and show students the jar.
3. Distribute the student page and allow students ample time to work through the problem.
4. When all groups have finished, invite a representative from each group to write their estimate on the board.
5. Discuss the estimates. Order them from low to high.
6. Invite each group to share how it determined the number of cube faces.
7. Find the actual number of cube faces.
8. Have the class critique the different strategies used.

Connecting Learning
1. What was your first reaction when you heard the problem?
2. Why did it seem like a huge problem?
3. How did you make it easier?
4. Which strategy did you think worked best?
5. What would be another problem like this that you would want to make easier?

* Reprinted with permission from *Principles and Standards for School Mathematics*, 2000 by the National Council of Teachers of Mathematics. All rights reserved.

Cube Face Estimation

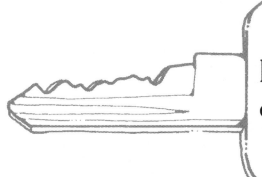

Key Question

How many faces are on the cubes in the estimation jar?

Learning Goal

Students will:

estimate the number of faces on the cubes in the estimation jar.

Cube Face Estimation

How many faces are on the cubes in the estimation jar?

Use the back of the paper to show how you solved this problem.

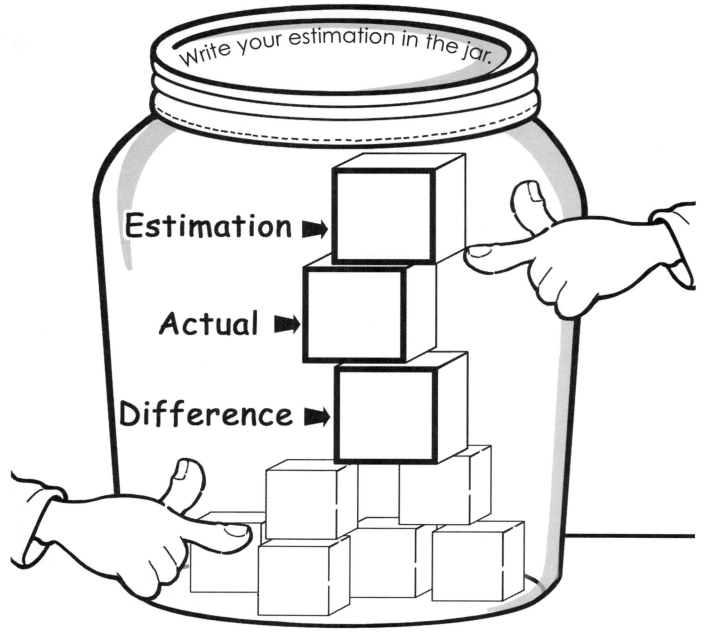

Write your estimation in the jar.

Estimation ▶

Actual ▶

Difference ▶

202

Cube Face Estimation

Connecting Learning

1. What was your first reaction when you heard the problem?

2. Why did it seem like a huge problem?

3. How did you make it easier?

4. Which strategy did you think worked best?

5. What would be another problem like this that you would want to make easier?

Practice Problems

The problems on the following pages are provided for additional practice with the problem-solving strategies covered in this book. No strategies have been recommended for the individual problems, and they do not follow any particular order. Students must decide which strategy to use based on the individual problem. It is suggested that the problems be copied onto transparencies and cut apart. A problem can then be placed on the overhead as a "bright beginning" to start math class or at any time during the day when a few minutes are available for review. To receive maximum benefit from the problems, be sure to have a time of discussion after each one where the emphasis is on the process and strategies used rather than arriving at the correct answer.

If you have to use 15 or fewer coins, how many different combinations of coins can be used to make $1.00?

How many ways can you show $\frac{1}{2}$ on a geoboard?

How many different ways can you group 12 beans?

Write the last 4 digits of your phone number on small pieces of paper. What is the largest 4-digit number you can make? ...the smallest?

How many different patterns can you make with 1 red, 1 yellow, 1 blue, and 1 white Unifix cube?

Use jumbo paper clips to find the perimeter of your notebook paper.

Megan sold 3 caramel apples to Ben, 5 to Akia, and 7 to Krystal. She had 4 left over. How many apples did she start with?

How can you use Teddy Bear Counters to show if a number is odd? ...even?

Complete this
in and out:

in	out
1	2
2	6
3	12
6	

5, 10, 15, 20…

What is the 27th number in the sequence?

There are 52 cookies to be arranged equally on 2 trays. How many cookies will there be on each tray?

2, 4, 6, 8…

What is the 17th number in the sequence?

Three alligators went down to the river and laid 5 eggs each. How many eggs were there in all?

How many toothpicks would be needed to make one row of 10 connected squares?

Ali bought a pumpkin from Sandia for $2.67. Ali gave Sandia $10.00. How much money will he get back?

How many dots would be needed for a 20-cube tower if you were going to place one dot on each exposed face?

There are 5 kids wearing shirts and pants. Four have white shirts. Two have white pants. How many shirts and pants are not white?

There are 99 roses. Each rose has 6 thorns. How many thorns are there on all the roses?

Four elephants go to the river. Three get their trunks wet. Two get their front feet wet. How many dry trunks and feet are there?

Laura has 3 green chips, 4 blue chips, and 1 red chip in her bag. What fractional part of the bag of chips is green?

Jerome walked 8 steps to the north, 6 steps to the east, 8 steps to the south, and 6 steps to the west. What number did he make?

There are 12 red and blue balls. Eight of them are red. How many balls are blue?

For Christmas, Mr. Lee gave $93 to his two sons. He gave $51 to his older son. How many dollars did he give the younger son?

There are 5 animals in the farmyard. There are 14 legs total. How many are chickens and how many are cows?

Flip a penny 20 times. Record the number of times it lands heads up and the number of times it lands tails up.

How many students in your class bring lunch? How many get lunch from the cafeteria? Survey your class and graph the results.

How many different ways can William make change for 50 cents without using pennies? Make a chart to show all possible ways.

How many planets have both rings and moons? Make a Venn diagram to compare all 9 planets.

How many ways can you make a sum of 12 using 2 whole numbers? How do you know you have all possible combinations?

How many ways can you make 15 cents using pennies, nickels, and dimes?

The school store sells erasers for 5¢, pencils for 10¢, paper for 20¢, and folders for 15¢. What combinations of supplies can Sean buy if he has 50¢?

How many capital letters have a horizontal line of symmetry?

Continue the pattern. Fill in the missing numbers:
7, 14, 21, ___, ___, ___

Pizza Place has 4 toppings: cheese, pepperoni, olives, and sausage. How many different 2-topping pizzas can they make?

Two numbers have a sum of 87. The larger of the numbers is twice the smaller. What are the numbers?

Continue the pattern. Fill in the missing numbers:
1, 2, 4, 7, 11, 16, ___, ___, ___

Fill in the missing days: Monday, Wednesday, Tuesday, Thursday, _____, _____, Thursday, Saturday

I have 7 coins with a total value of $0.57. What coins do I have? How many of each coin do I have?

Bradley has 2 cousins. The sum of their ages is 20. One cousin is 4 years older than the other. What are their ages?

I have 4 coins. Two of my coins are the same. The total value of my coins is 36 cents. What coins do I have?

It takes 4 toothpicks to make a square. It takes 8 toothpicks to make 2 squares. How many toothpicks will it take to make 3 squares? ...5 squares? ...34 squares?

There are 2 more bats than rats in the haunted house. The number of bats is half the number of cats. There are 14 animals total. How many are rats, bats, and cats?

Markesha is taking cold medicine. She has to take 1 teaspoon every 45 minutes. She took her first dose at 2:20. Make a chart to show her medicine schedule. Can she get 6 doses in before her bedtime at 8:00?

Jake has 2 dogs at home. Julie has 4 cats at home. Sue has 3 cats and 1 dog. Chhay has 3 fish. Make a graph that shows how many of each of the animals the friends have.

Emanuel has some gum balls. Joan had 3 times as many as Emanuel. She ate 4 and now she has 5. How many gum balls does Emanuel have?

It takes Jeremy 15 minutes to walk to school. It takes him 30 minutes to get ready. Breakfast takes 15 minutes. School starts at 8:30. When should Jeremy get up to be on time?

Measure the length of your desk in jumbo paper clips. Two jumbo paper clips are equal to 3 small paper clips. How many small paper clips long is your desk?

If I get a dime on Monday and double my money on Tuesday, Wednesday, Thursday, and Friday, how much money will I have on Saturday?

The parking lot at the amusement park can hold 220 cars. Only 5 cars can be parked in each row. How many rows are there?

Chris is training his sister to climb stairs. She can climb up 1 or 2 stairs at a time. If a flight of stairs has 10 steps, how many ways can his sister climb up the flight of stairs?

Ms. Woo has a small library at home. She has 319 books on history, 51 books on geography, and 192 books on science. How many books does she have in all?

A train ticket to Bend costs 4 times as much as a ticket to Sandy Beach. If the ticket to Bend costs $48, how many dollars is the ticket to Sandy Beach?

Chin took Mr. Brown up 6 floors from the floor he lives on. Then Chin went down 5 floors, where he picked up Mrs. Lee. Chin took her down 10 floors to the first floor lobby. What is the number of the floor Mr. Brown lives on?

Channel 4 wants to show wildlife films from 1:00 P.M. to 5:00 P.M. on Thursday. If each program lasts half an hour, how many different programs will they need to buy to fill the time slot?

Marcus had some marbles. He bought 42 more from the grocery store, 15 more from the department store and found 47 more on the playground. Now he has 184 marbles. How many marbles did he start with?

Kayley had a total of 10 rocks in both of her pockets. She wanted to have an equal number of rocks in each of her pockets, so she moved 2 rocks from her left pocket to her right pocket. How many rocks did she have in her left pocket before?

The class roll only has last names: Green, Brown, and Smith. Name tags give only first names: Kim, Brad, and Jamal. Match the first and last names. Brad's last name is not Smith. Kim and Brad are cousins. Jamal's initials are JLB.

Anne has 3 banks where she keeps her money. In the first bank there is $7. In the second bank there is $3 more than in the first. In the third bank there is $2 less than in the second bank. How much money does she have altogether?

There were three lunch boxes. One was blue, one was red, and one was yellow. One box had a pear, one had a banana, and one had an orange. The box with the pear was not red. The blue box contained a banana. Which fruit was in which box?

Nancy, Shereka and Mary were Olympic divers. They took 1st, 2nd, and 3rd place in diving. Use these clues to put the divers in order: The first place diver was not Mary. Nancy took third place. Shereka wore blue.

Four friends are standing in the lunch line. Bill is in front of John. Christian is behind Bo. John and Bo are in the middle of the line. In what order are the friends standing?

Every space in the parking lot was filled. Kareem's car was in the middle. There were 6 cars to the right of his car. How many cars were in the parking lot?

There are 4 boys in the Mitchell family. David is older than Nick and younger than Charles. Andrew is not the oldest or the youngest. David does not have 2 older brothers. List the names of the boys from oldest to youngest.

Five colored blocks made a tower. One block is red, two are blue, and two are green. No two blocks of the same color are together. The top block is not green. The center block is not red. One green block is just below the red one. The bottom block is not green.

A group of basketball players is standing in a circle. Everyone faces someone across the circle. The players count off in order starting with number 1. Player 2 is directly across from player 7. How many players are in the circle?

Kyle wanted to get to school 20 minutes early. School starts at half-past eight. He got there at a quarter after eight. Was he as early as he wanted to be?

Celine wants to watch her favorite TV show at 7:00. It takes her 20 minutes to study her spelling words, 15 minutes to practice her multiplication facts, and 15 minutes to do her chores. What time should she start so that she will be finished in time to watch her show?

How many blocks would be needed for a 5-step staircase that starts like this?

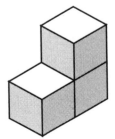

The AIMS Program

AIMS is the acronym for "Activities Integrating Mathematics and Science." Such integration enriches learning and makes it meaningful and holistic. AIMS began as a project of Fresno Pacific University to integrate the study of mathematics and science in grades K-9, but has since expanded to include language arts, social studies, and other disciplines.

AIMS is a continuing program of the non-profit AIMS Education Foundation. It had its inception in a National Science Foundation funded program whose purpose was to explore the effectiveness of integrating mathematics and science. The project directors in cooperation with 80 elementary classroom teachers devoted two years to a thorough field-testing of the results and implications of integration.

The approach met with such positive results that the decision was made to launch a program to create instructional materials incorporating this concept. Despite the fact that thoughtful educators have long recommended an integrative approach, very little appropriate material was available in 1981 when the project began. A series of writing projects have ensued, and today the AIMS Education Foundation is committed to continue the creation of new integrated activities on a permanent basis.

The AIMS program is funded through the sale of books, products, and staff development workshops and through proceeds from the Foundation's endowment. All net income from program and products flows into a trust fund administered by the AIMS Education Foundation. Use of these funds is restricted to support of research, development, and publication of new materials. Writers donate all their rights to the Foundation to support its on-going program. No royalties are paid to the writers.

The rationale for integration lies in the fact that science, mathematics, language arts, social studies, etc., are integrally interwoven in the real world from which it follows that they should be similarly treated in the classroom where we are preparing students to live in that world. Teachers who use the AIMS program give enthusiastic endorsement to the effectiveness of this approach.

Science encompasses the art of questioning, investigating, hypothesizing, discovering, and communicating. Mathematics is the language that provides clarity, objectivity, and understanding. The language arts provide us powerful tools of communication. Many of the major contemporary societal issues stem from advancements in science and must be studied in the context of the social sciences. Therefore, it is timely that all of us take seriously a more holistic mode of educating our students. This goal motivates all who are associated with the AIMS Program. We invite you to join us in this effort.

Meaningful integration of knowledge is a major recommendation coming from the nation's professional science and mathematics associations. The American Association for the Advancement of Science in *Science for All Americans* strongly recommends the integration of mathematics, science, and technology. The National Council of Teachers of Mathematics places strong emphasis on applications of mathematics such as are found in science investigations. AIMS is fully aligned with these recommendations.

Extensive field testing of AIMS investigations confirms these beneficial results:

1. Mathematics becomes more meaningful, hence more useful, when it is applied to situations that interest students.
2. The extent to which science is studied and understood is increased, with a significant economy of time, when mathematics and science are integrated.
3. There is improved quality of learning and retention, supporting the thesis that learning that is meaningful and relevant is more effective.
4. Motivation and involvement are increased dramatically as students investigate real-world situations and participate actively in the process.

We invite you to become part of this classroom teacher movement by using an integrated approach to learning and sharing any suggestions you may have. The AIMS Program welcomes you!

AIMS Education Foundation Programs

Practical proven strategies to improve student achievement

When you host an AIMS workshop for elementary and middle school educators, you will know your teachers are receiving effective usable training they can apply in their classrooms immediately.

Designed for teachers—AIMS Workshops:
- Correlate to your state standards;
- Address key topic areas, including math content, science content, problem solving, and process skills;
- Teach you how to use AIMS' effective hands-on approach;
- Provide practice of activity-based teaching;
- Address classroom management issues, higher-order thinking skills, and materials;
- Give you AIMS resources; and
- Offer college (graduate-level) credits for many courses.

Aligned to district and administrator needs—AIMS workshops offer:
- Flexible scheduling and grade span options;
- Custom (one-, two-, or three-day) workshops to meet specific schedule, topic and grade-span needs;
- Pre-packaged one-day workshops on most major topics—only $3900 for up to 30 participants (includes all materials and expenses);
- Prepackaged four- or five-day workshops for in-depth math and science training—only $12,300 for up to 30 participants (includes all materials and expenses);
- Sustained staff development, by scheduling workshops throughout the school year and including follow-up and assessment;
- Eligibility for funding under the Title I and Title II sections of No Child Left Behind; and
- Affordable professional development—save when you schedule consecutive-day workshops.

University Credit—Correspondence Courses

AIMS offers correspondence courses through a partnership with Fresno Pacific University.
- Convenient distance-learning courses—you study at your own pace and schedule. No computer or Internet access required!

The tuition for each three-semester unit graduate-level course is $264 plus a materials fee.

The AIMS Instructional Leadership Program

This is an AIMS staff-development program seeking to prepare facilitators for leadership roles in science/math education in their home districts or regions. Upon successful completion of the program, trained facilitators become members of the AIMS Instructional Leadership Network, qualified to conduct AIMS workshops, teach AIMS in-service courses for college credit, and serve as AIMS consultants. Intensive training is provided in mathematics, science, process and thinking skills, workshop management, and other relevant topics.

Introducing AIMS Science Core Curriculum

Developed to meet 100% of your state's standards, AIMS' Science Core Curriculum gives students the opportunity to build content knowledge, thinking skills, and fundamental science processes.
- *Each* grade specific module has been developed to extend the AIMS approach to full-year science programs.
- *Each* standards-based module includes math, reading, hands-on investigations, and assessments.

Like all AIMS resources, these core modules are able to serve students at all stages of readiness, making these a great value across the grades served in your school.

For current information regarding the programs described above, please complete the following form and mail it to: P.O. Box 8120, Fresno, CA 93747.

Information Request

Please send current information on the items checked:

____ *Basic Information Packet* on AIMS materials ____ Hosting information for AIMS workshops
____ *AIMS Instructional Leadership Program* ____ AIMS Science Core Curriculum

Name _____ Phone _____

Address_____
       Street   City   State   Zip

Magazine

YOUR K-9 MATH AND SCIENCE
CLASSROOM ACTIVITIES RESOURCE

The AIMS Magazine is your source for standards-based, hands-on math and science investigations. Each issue is filled with teacher-friendly, ready-to-use activities that engage students in meaningful learning.

• *Four issues each year (fall, winter, spring, and summer).*

Current issue is shipped with all past issues within that volume.

| 1820 | Volume XX | 2005-2006 | $19.95 |
| 1821 | Volume XXI | 2006-2007 | $19.95 |

Two-Volume Combination
| M20507 | Volumes XX & XXI | 2005-2007 | $34.95 |

Back Volumes Available
Complete volumes available for purchase:

1802	Volume II	1987-1988	$19.95
1804	Volume IV	1989-1990	$19.95
1805	Volume V	1990-1991	$19.95
1807	Volume VII	1992-1993	$19.95
1808	Volume VIII	1993-1994	$19.95
1809	Volume IX	1994-1995	$19.95
1810	Volume X	1995-1996	$19.95
1811	Volume XI	1996-1997	$19.95
1812	Volume XII	1997-1998	$19.95
1813	Volume XIII	1998-1999	$19.95
1814	Volume XIV	1999-2000	$19.95
1815	Volume XV	2000-2001	$19.95
1816	Volume XVI	2001-2002	$19.95
1817	Volume XVII	2002-2003	$19.95
1818	Volume XVIII	2003-2004	$19.95
1819	Volume XIX	2004-2005	$35.00

Call today to order back volumes: 1.888.733.2467.

**Call 1.888.733.2467 or
go to www.aimsedu.org**

Subscribe to the AIMS Magazine

$19.95 a year!

AIMS Magazine is published four times a year.

Subscriptions ordered at any time will receive all the issues for that year.

AIMS Online – www.aimsedu.org

For the latest on AIMS publications, tips, information, and promotional offers, check out AIMS on the web at www.aimsedu.org. Explore our activities, database, discover featured activities, and get information on our college courses and workshops, too.

AIMS News

While visiting the AIMS website, sign up for AIMS News, our FREE e-mail newsletter. Published semi-monthly, AIMS News brings you food for thought and subscriber-only savings and specials. Each issue delivers:

• **Thought-provoking articles on curriculum and pedagogy;**
• **Information about our newest books and products; and**
• **Sample activities.**

Sign up today!

AIMS Program Publications

Actions with Fractions, 4-9
Awesome Addition and Super Subtraction, 2-3
Bats Incredible! 2-4
Brick Layers II, 4-9
Chemistry Matters, 4-7
Counting on Coins, K-2
Cycles of Knowing and Growing, 1-3
Crazy about Cotton, 3-7
Critters, 2-5
Electrical Connections, 4-9
Exploring Environments, K-6
Fabulous Fractions, 3-6
Fall into Math and Science, K-1
Field Detectives, 3-6
Finding Your Bearings, 4-9
Floaters and Sinkers, 5-9
From Head to Toe, 5-9
Fun with Foods, 5-9
Glide into Winter with Math and Science, K-1
Gravity Rules! 5-12
Hardhatting in a Geo-World, 3-5
It's About Time, K-2
It Must Be A Bird, Pre-K-2
Jaw Breakers and Heart Thumpers, 3-5
Looking at Geometry, 6-9
Looking at Lines, 6-9
Machine Shop, 5-9
Magnificent Microworld Adventures, 5-9
Marvelous Multiplication and Dazzling Division, 4-5
Math + Science, A Solution, 5-9
Mostly Magnets, 2-8
Movie Math Mania, 6-9
Multiplication the Algebra Way, 4-8
Off the Wall Science, 3-9
Out of This World, 4-8
Paper Square Geometry:
    The Mathematics of Origami, 5-12
Puzzle Play, 4-8
Pieces and Patterns, 5-9
Popping With Power, 3-5
Positive vs. Negative, 6-9
Primarily Bears, K-6
Primarily Earth, K-3
Primarily Physics, K-3
Primarily Plants, K-3

Problem Solving: Just for the Fun of It! 4-9
Problem Solving: Just for the Fun of It! Book Two, 4-9
Proportional Reasoning, 6-9
Ray's Reflections, 4-8
Sense-Able Science, K-1
Soap Films and Bubbles, 4-9
Solve It! K-1: Problem-Solving Strategies, K-1
Solve It! 2nd: Problem-Solving Strategies, 2
Solve It! 3rd: Problem-Solving Strategies, 3
Spatial Visualization, 4-9
Spills and Ripples, 5-12
Spring into Math and Science, K-1
The Amazing Circle, 4-9
The Budding Botanist, 3-6
The Sky's the Limit, 5-9
Through the Eyes of the Explorers, 5-9
Under Construction, K-2
Water Precious Water, 2-6
Weather Sense: Temperature, Air Pressure, and Wind, 4-5
Weather Sense: Moisture, 4-5
Winter Wonders, K-2

Spanish Supplements*
Fall Into Math and Science, K-1
Glide Into Winter with Math and Science, K-1
Mostly Magnets, 2-8
Pieces and Patterns, 5-9
Primarily Bears, K-6
Primarily Physics, K-3
Sense-Able Science, K-1
Spring Into Math and Science, K-1

* Spanish supplements are only available as downloads from the
   AIMS website. The supplements contain only the student pages
   in Spanish; you will need the English version of the book for the
   teacher's text.

Spanish Edition
Constructores II: Ingeniería Creativa Con Construcciones
    LEGO® 4-9
    The entire book is written in Spanish. English pages not included.

Other Science and Math Publications
Historical Connections in Mathematics, Vol. I, 5-9
Historical Connections in Mathematics, Vol. II, 5-9
Historical Connections in Mathematics, Vol. III, 5-9
Mathematicians are People, Too
Mathematicians are People, Too, Vol. II
What's Next, Volume 1, 4-12
What's Next, Volume 2, 4-12
What's Next, Volume 3, 4-12

For further information write to:
AIMS Education Foundation • P.O. Box 8120 • Fresno, California 93747-8120
www.aimsedu.org • 559.255.6396 (fax) • 888.733.2467 (toll free)

Duplication Rights

Standard Duplication Rights

Purchasers of AIMS activities (individually or in books and magazines) may make up to 200 copies of any portion of the purchased activities, provided these copies will be used for educational purposes and only at one school site.

Workshop or conference presenters may make one copy of a purchased activity for each participant, with a limit of five activities per workshop or conference session.

Standard duplication rights apply to activities received at workshops, free sample activities provided by AIMS, and activities received by conference participants.

All copies must bear the AIMS Education Foundation copyright information.

Unlimited Duplication Rights

To ensure compliance with copyright regulations, AIMS users may upgrade from standard to unlimited duplication rights. Such rights permit unlimited duplication of purchased activities (including revisions) for use at a given school site.

Activities received at workshops are eligible for upgrade from standard to unlimited duplication rights.

Free sample activities and activities received as a conference participant are not eligible for upgrade from standard to unlimited duplication rights.

Upgrade Fees

The fees for upgrading from standard to unlimited duplication rights are:
- $5 per activity per site,
- $25 per book per site, and
- $10 per magazine issue per site.

The cost of upgrading is shown in the following examples:
- activity: 5 activities x 5 sites x $5 = $125
- book: 10 books x 5 sites x $25 = $1250
- magazine issue: 1 issue x 5 sites x $10 = $50

Purchasing Unlimited Duplication Rights

To purchase unlimited duplication rights, please provide us the following:
1. The name of the individual responsible for coordinating the purchase of duplication rights.
2. The title of each book, activity, and magazine issue to be covered.
3. The number of school sites and name of each site for which rights are being purchased.
4. Payment (check, purchase order, credit card)

Requested duplication rights are automatically authorized with payment. The individual responsible for coordinating the purchase of duplication rights will be sent a certificate verifying the purchase.

Internet Use

Permission to make AIMS activities available on the Internet is determined on a case-by-case basis.

• P. O. Box 8120, Fresno, CA 93747-8120 •
• permissions@aimsedu.org • www.aimsedu.org •
• 559.255.6396 (fax) • 888.733.2467 (toll free) •